"As a coach, speaker and author focuse
ness, I am excited to share *Boomerang* \
moving forward.

"In this great resource, Tyler Smith pulls back the curtain to share sound strategies for retaining first-time and multi-visit guests at your church. Filled with real world stories and examples, his modern approach to making the most of the guest experience, is a fresh playbook that is chock full of ideas that your church can implement.

"In a world where mobile devices have practically become an extra appendage, Smith helps us make important shifts in guest follow-up that are essential to doing ministry in this present time. This gold mine of great ideas are accessible for churches of all sizes, styles and means. And no, you don't have to have oodles of money, or be a tech wiz to make these learning a reality in your church.

"I'll be adding this to my seminar handbook as a recommended resource!"

—Jason Moore
Author | Speaker | Worship and Guest Readiness Coach

"At Church Marketing University we work with thousands of churches across the globe. The vast majority of those churches struggle with getting guests to return especially after big events, outreaches, Christmas, and Easter. And that is why we love Text In Church and their *Boomerang* framework. It's a system that works for moving your church beyond the big event to big long-term impact."

—Ryan Wakefield
Church Marketing University

"Our first-time guest engagement has drastically increased. People appreciate the personalized follow-up messages. And the automated workflow make it simple and easy! It has also provided another option for new believers to respond (via text at the end of service) and engage with our staff in next steps. Also very useful for group SMS messages to remind our people about events and meetings."

—David Bauchspiess, River Church

"While reading *Boomerang*, I was bummed because I didn't have this when starting Churches in the past. I love practical/achievable steps and this book gives you exactly that to reach more people with the Gospel. You need this book!"

—Zachary Minton, The Rock Church

"Effective communication is essential to ministry success. Text In Church has revolutionized the way we communicate in our church in outreach, in-reach, and volunteer sign ups resulting in both growth and greater involvement."

—Pastor Zac Clay, Pine Grove Baptist Church

"If you've been on the fence thinking there is NO easy way to increase guest retention in your church, think again. *Boomerang* will blow your mind, and on the path to beating national averages with little time investment"

—Kevin Ranfeld, Family Church

"We expect a *Boomerang* to return. Now let's EXPECT guests to return to our churches. We need to be known for it! Leave it to the Text In Church experts to create a simple system and strategy. Read on!"

—Mark MacDonald
Sr. Church Brand Consultant and Bestselling Author
Be Known For Something

"Nobody falling through the cracks again!!! From reach to retention!!!"

—Nathan Johnson, Unite 2 Ignite

"Tyler and Ali want to see every church succeed, that's why he's giving away the secret sauce for guest retention in *Boomerang*. If you want to close the back door at your church and help assimilate a higher percentage of first time guests, then you need to read this!"

—Ben Stapley, Christ Fellowship

"Thanks to the Text In Church team for giving us those templates and for giving us some language to use to have people come back. We now have a follow-up process to invite people that visited just one time and we have seen huge growth inside our church."

—Drew Keller, Elevate Church

"The positive feedback I get on a constant basis is outstanding, and people love how they are thought about and connected with after they leave our service."

—Danielle Krivda, Seventh-day Adventist Church

"I love how easy it is to customize the follow-up steps for our visitors. We have had great results after only 2 months of use."

—Ladona Smith, Second Baptist Church

"This system has helped bring people in the door, get them back each week and keep them informed of future events coming up. It's a great program that has helped our church reach our community."

—Chris Rathbone, The City Church

"Our personal connection to first time guests has increased tremendously . . . and thus our conversion of guests to members has increased as well. Amazing program . . . and super easy to use."

—Michael Pippin, The Crossing Church

"Since implementing our plans, we have seen a surge in guest retention and a 27% increase in overall attendance."

—Chris Sykes, The Church at Lake Forest

"Great tool to help with the follow-up process. Text In Church offers templates to use and makes it so easy to reach out to visitors. Offering the what to do and when to do it takes the guess work out of what really works."

—Angie Shaw, White Marsh Baptist Church

"These guys have an intense focus on seeing churches build systems for following up with the people they encounter. They are paving the way for many churches to be able to connect with new people on a personal level. Thank you so much!"

—Mark Tenney, Digital Church Platform

"The automated follow-up campaign helps prevent guests from slipping through the cracks and ensures that we are following up with them for 6 weeks!"

—Derrick Abell, Fellowship of Grace

"I was hesitant—thinking our church was not big enough for this yet- but I was wrong Text In Church has helped us build our connections with new people, saved us time with automated workflows, and it has been way more effective than the plan emails we were using. Honestly, everyone is going to check a text they got. It has enabled us to develop relationships with more people than what we were doing before. Give it a try, you really don't have anything to lose. On the plus side, the support is incredible. There is no dumb question, and they'll make sure you're good to go! (trust me, I've asked a ton of questions)."

—Billy Arreola Lopez, Christ Community Church

"This process works. In 6 months we went from 10% to 56% new guests not only returning, but becoming members, regular attenders or taking next steps."

—Faith Brown, Crosspoint Fellowship Church

"These guys have filled a gap—they make SMS text communication with our congregation easy, especially for guest follow-up. And they KNOW guest follow-up—began using their starter templates immediately. Just great stuff."

—Patrick Bradley, Passion 4 Planting

"It's been awesome for me to interact with more people. But, it's also give us an opportunity to get great feedback from our guests as well."

—Kenny Kirby, Mountain View Community Church

BOOMERANG

THE POWER OF EFFECTIVE GUEST FOLLOW-UP

**TYLER SMITH &
ALISON HOFMEYER**

Cover Design: Jenna Vreugdenhil
Illustrations: Rachel O'Brien
Interior Design and Typesetting: Mandi Cofer, thetinytypesetter.com

Printed in the United States of America

BOOMERANG IS DEDICATED TO:

The entire Text In Church staff. All of our company's success is wrapped up in the individual skills and passions of each of you. We would not exist as the incredible system that we are without each other and I am so grateful for each of you.

My family, who has championed my "crazy ideas" and motivation to always push harder. Without your support, none of this could be. You carry me and I love you.

The local church. I can't imagine a world without you. You're living out the love of Jesus and I'm so honored every day to work alongside you.

Text In Church is the communication platform for churches, hosting email and text messaging capabilities on a web-based platform. Text In Church works with over 17,000 church leaders and is growing every year. The heart of the company is to connect churches with their communities and free church leaders up to spend more time doing what got them into ministry in the first place.

Boomerang is the culmination of everything we have learned from working with over 17,000 church leaders. All of the strategies and heart behind the who, what, when, where, why, and how of guest follow-up is laid out in this book in hopes that it will serve as a very practical, daily guide. A boomerang comes back. But, only when thrown correctly. The guest experience is happening at your church, it's happening at every church in our world today. However, that doesn't mean it's a positive thing, and it definitely doesn't always cause guests to want to come back. What we've discovered, though, is a plan, a "throwing guide" if you will, that when pieced together with time and intentionality, will create a thoughtful, unforgettable experience that will keep your first time guests coming back.

CONTENTS

Contents

FOREWORD

by Jonathan Malm

acclaimed writer and entrepreneur

Tools are only as good as your plan to use them.

When I was asked to help create a plan to turn outsiders into insiders at my church, I was excited for the opportunity. We had all the tools we could need, but I knew there was a process that was much more important than the tools. My goal was to balance the impersonal process of tracking people with the personal touch of communicating with people. I didn't want things to feel impersonal. I didn't want the whole thing to feel like they were getting processed by a machine. And my pastor didn't want it to take 10 hours of staff time each week to make it happen. Text In Church has been a vital part of that process. It mixes accountability, sustainability, and the personal touch. We've noticed a huge response from the people we text. They feel like they're talking to a real person (because they are) and actually respond. That's why I was excited when Tyler Smith told me about this book. All of the content in this book exists to walk church leaders through this process of creating a system that will genuinely and effectively follow up with first time guests. It's a guide to a guest follow up plan that works. I know that because I've seen it at my own church. Not only that, I know I still have lots to learn through experimentation and learning best practices from others that have done this. My prayer for you is that you would take the truths in this book, contextualize them for your church's unique DNA, and see massive community growth at your church. Our world is only as strong as the communities in our churches, and I pray yours would be one of the best at connecting new people to the life at your church.

BOOMERANG

THE POWER OF EFFECTIVE GUEST FOLLOW-UP

BOOMERANG: THE PROVEN FOLLOW-UP SYSTEM THAT KEEPS YOUR CHURCH GUESTS COMING BACK

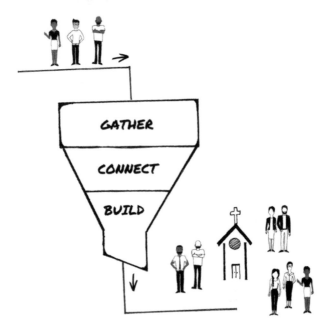

INTRODUCTION

We were a young congregation worshiping in a historic building. We had recently begun the church planting process and were still at the early stages. The stories our church building could tell were countless. It had functioned as a hospital during the Civil War, held more weddings and funerals than we could track, and was the home to a once-thriving church.

So there we were—the leadership team—trying our best to revive this sacred space and grow our faith community.

One winter night, our team gathered to discuss growth strategies. Specifically, we were re-evaluating our guest follow-up strategies. Our attendance had grown a bit, but we weren't able to sustain it. The six of us sat in a small, cold room, for what seemed like hours, mapping out every detail that should go into an effective guest follow-up process. We covered a whiteboard as we brainstormed ideas. In the end, we settled on about fourteen individual follow-up steps. We worked together, and everyone bought in and was even, maybe, dare I say, excited about our plan. As we wrapped up, our pastor asked a question that silenced the room.

"Okay, so who wants to tackle this?"

Crickets. No one raised their hand. It seemed too daunting. The time requirement was too much. (Sound familiar?)

All we wanted to do was connect more effectively with our first-time guests. But we didn't have the time or resources to do it.

On the way home that night, my phone buzzed with a text message. It was Southwest Airlines texting to tell me that my flight for the next morning was cancelled. This inconvenience, paired with the horrible end to our planning meeting, sent my mood over the edge. But, while I was wallowing in my frustration and disappointment, it hit me. If Southwest can leverage technology to get their message out, why can't the church?

That was about ten years ago, and that began my journey of discovering what tools churches need and how technology can provide those tools to empower churches to move from surviving to thriving. Much has been born out of this journey. The ENGAGE Conference hosts more than 21,000 church leaders annually at a free, all-online conference. Leaders in the church and technology spaces present over thirty-five sessions on how to utilize technology to grow your church. Boomerang is an educational tool for churches to use to walk, step by step, through the process of discovering and implementing the best guest follow-up strategy at their church. And Text In Church is a software that serves over 17,000 church leaders through its automation and messaging capabilities.

My purpose in writing this book is simple: I want to provide the most tactical and practical tools for churches to engage with their first-time guests.

Connecting with our congregations and communities shouldn't be this hard!

But it is. I have heard from pastors all across the country who say it, in fact, feels like an uphill battle to genuinely connect with all the people who walk through their doors, specifically with first-time guests.

We've been blown away by the effort church leaders, communicators, and volunteers have made to create these connections. Welcome teams work tirelessly to identify new people and make them feel welcome and celebrated. Worship services are thought through carefully and executed with so much intentionality. Sunday morning and midweek programming is planned months in advance and often has its own curriculum. Sermons are relevant and are the building blocks for the rest of the content presented during the week at church or online.

And, of course, your worship, preaching, and Sunday school classes are all *super* important. Your church needs all of those moving parts to operate smoothly, and these might be part of what keeps people at your church for the long haul. But that's not why people are initially showing up. You see, most people, when they come to your church for the first time, are showing up for one of five reasons. This is known as the "5-D Theory":

THE 5-D THEORY

1. Divorce
2. Death in the family
3. Displacement (a move)
4. Disaster
5. Development (for self, children, marriage, etc.)

They're showing up because they're real people with real lives, real hurts, and real questions. And they're searching.

When someone looks up your church online, it's probably not because they suddenly think, *Hmm, maybe I'll try a brand-new church tomorrow!*

The truth is, it's more likely that they're desperate. They're lonely. Maybe they're looking for some hope or just some direction as they navigate a new stage of life.

And, well . . . church just might be the last resort.

So, our response to this should be relatively simple. Connect with them. Throw them a lifeline. Make them feel loved.

From the moment people check out your church online, you have an opportunity to make an impact on them. Show them that you're a real person, that the whole church is made up of real people, and that you want to invite them to be a part of your community! The truth is, people aren't looking for a friendly church; they are looking for a friend.

Connection isn't complicated, but the execution requires two important things: time and resources—two things we never seem to have enough of.

But what if I told you that I could give you some of your time back? What if I told you that you could, in fact, improve communications at your church without spending more time managing data, adding more technology, and drafting emails. At Text In Church, we've landed on a solid, simple strategy, and that's what I'm going to lay out for you in this book. I'll show you all the things your church could be doing, including using technology, to effectively connect with guests and members. And I'll show you how to do all of this in a way that actually frees up more of your time to spend doing the things that got you into ministry in the first place.

Leveraging technology to forge relationships has been transformational for me and thousands of churches across the country. If you call me anything, call me a good listener. I've listened to the needs of churches and built the technology to support those needs. It's nothing complicated or uniquely mine. In fact, the inspiration for all of it comes from the church leaders I work with on a daily basis; they're the real secret sauce to all of this. And now we are so thankful for the opportunity to share it with you.

The first church I worked with was my brother's. He's a pastor and church planter in California. While he was incredibly skeptical of the idea at first—I think the words "terrible" and "spammy" were how he described it—he became my guinea pig. His skepticism was valid. I mean, an airline communicating with its passengers is vastly different than church leaders seeking to build authentic relationships with those in their community. But he committed to figuring it out with me. I asked him, "If I was a full-time employee at your church, hired only for follow-up, what would you have me do?"

He gave me a "job description," and I took that and automated what I could through texts and emails. And, much to both of our surprise, it worked.

What started out as a way to help my brother's and my own church grow turned into a full-blown follow-up system for first-time guests. It sounds like a lot, but it can be accomplished in just three steps:

1. Gather
2. Connect
3. Build

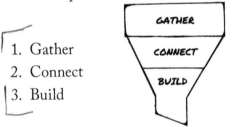

These three steps are the framework for this book—and for the strategy that has helped thousands of churches all across the country see significant growth.

Take Kevin, for example. He's a pastor at a small church in Missouri. He listened to one of our webinars on these three steps and loved the concept. Worried his church wouldn't be able to purchase our tool, he decided just to take the strategies we teach and implement them himself. In the first three to four months of doing this, he saw a 25 percent increase in guest retention.

After realizing that these strategies do work and spending too many hours per week doing the follow-up himself, he decided to purchase our platform. He then saw another 28 percent increase in guest retention! That's a 53 percent increase in guest retention over the short term. Their long-term guest retention (over twelve months) also saw a rise, growing to 30 percent (almost double the national average)!

There's also Faith. She runs the First Impressions ministry at a mid-sized family church in Texas. They were seeing an increase in first-time guests after implementing Plan a Visit on their church website. (Don't worry. We will talk through Plan a Visit in great detail in Chapter 2.) Approximately thirteen new guests visited the church weekly, and around 16 percent were taking next steps. While they were pleased with the influx of first-time guests, they wanted more than just a bunch of visitors. They wanted these first-time guests to truly feel connected and to come back. So Faith developed a six-week follow-up plan herself that consisted of texts and emails. Her "ah-ha!" moment was on Easter Sunday. Their church saw seventeen first-time guests that day. Faith spent two and a half hours creating the personalized messages to go out to each of these people. She was creating over seventy weekly emails and text messages. Did I mention she's also a mom of 6 kids, a wife, and only works part time for her church? She recognized that this was not scalable. Their church got Text In Church, and Faith was able to see all of her hard work come to fruition. Not only was her time freed up—I think she spent just a few hours setting up her automated workflows to first-time guests (Yes, we will get into those later, too.)—but she was also finally seeing the growth she knew was coming because the process didn't cap out with her. Their church is seeing an average of 123 more people in attendance on Sunday morning than they did at the same time last year. And, 67 percent of their first-time guests are taking next steps!

These are just a couple of the thousands of success stories Text In Church has seen. These steps work. And they work whether or not you buy the tool we've created to do it. The tool saves you time, but the strategies make the difference. And in the pages that follow, I am going to walk you through every single one of them.

But before we dive in, there is a specific posture that I think is critical. It has made the difference between churches spinning their wheels and overworking and churches implementing this plan with ease. That is:

- First-time guests need to be obsessed over.
- First-time guests need to walk through the doors of your church and think, *They were hoping I would be here!*
- They need to leave your service and think, *They hope I come back.*

- They need to be followed up with and left thinking, *They care about me.*

If a guest feels invisible when they are at your church, no amount of follow-up is going to undo that. I have a lot more to say about being obsessed with first-time guests. I also have a free resource we'd like to offer you on how to plan your services with first-time guests in mind. We will talk more about both in the coming chapters, but you can also access this content in our free Boomerang Kit. You get the kit free with the purchase of this book. Just text "kit" to 816-482-3337, and we will send you the link!

If there's only one thing you glean from this book, let it be this: you can save time and connect better with first-time guests. Yes, you read that right. I want to save you time and help you connect with more people more often. This isn't an impossible task, not by any stretch of the imagination. The better churches get at connecting with first-time guests, the more guests come back to church. The more people in church, the further the Gospel is spread. That's the Great Commission in action. Boom!

Churches in America are dying. But that's not because the message has changed. People, technology, and culture have all changed. We have to change, too, by caring, communicating, and connecting with people after they leave the doors of our church. We have seen churches experience incredible growth with the use of our system, and we are excited to lay the playbook wide open for you!

GATHER

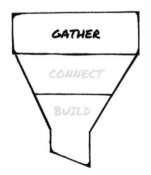

I visited a church with my family that was so friendly! We probably met ten new people our first Sunday there! We were definitely noticed, and people were excited we were there. I wish that would have been enough. However, we didn't have anything tangible. No next step. No contact person. And we didn't leave our info with anyone. So, that was it. A great Sunday, for sure! But it ended there; we never went back.

—Anonymous Visitor

G ather is the first step in our three-step process. While the concept is simple—gather someone's contact information—the approach can be multifaceted. The gather step is also foundational to the other two steps. All of the incredible connection and relationship-building you can start fostering outside of your worship experience is possible when you have contact information for people. Unfortunately, secular culture has gotten really good at telling people how many other things they should do during the week besides go to church. But when people do go, and they start recognizing faces, maybe share a meal with someone, and really get connected, they realize that church is much more than what happens for an hour on a Sunday morning. So don't just think about this as that hard hurdle to get over with first-time guests. Think of it as an investment, a crucial first step in building genuine, long-term relationships with the people in your community. An investment definitely worth making.

HUGE

CHAPTER 1

THE GUEST-OBSESSED WEBSITE

Do you know the first place people go to when looking for a church?

Google.

This is probably no surprise. But what might surprise you is that **seventeen million people who are *not* regular churchgoers visit church websites each year**. Whether they were invited by a friend or simply looking on their own, they all start in the same place. Online.

And that's where you must start too. With your church website. And more specifically, with creating a guest-obsessed website.

Connecting with the unchurched through your website is an incredible opportunity and responsibility. This doesn't mean your church website has to be expensive or custom-made by a developer. It's better to have less content that is focused on answering the questions people have when they visit your site than to have pages of content that are confusing to navigate. In this case, less is more. Website design and creation is not my niche, so I'll leave those nitty-gritty details to the professionals. However, there are four crucial components that must appear on your church's website if you want it to engage with potential first-time guests:

1. Service Times and Location
2. About Us Page
3. Imagery
4. Plan A Visit Feature
   (We'll talk about this more in a separate chapter.)

Let's dive in to each one of these components.

SERVICE TIMES AND LOCATION

Have you ever wondered what time Target closes? Or the hours for that local restaurant you've been wanting to try? How do you find that information? You Google it. The best part is, the hours show up right on the search page. You don't have to dig through the website to find out when they're open. The same should be true for churches! People need to know when and where to go, and they shouldn't have to search long for that information. It needs to be highly visible on the homepage of your website.

> *I learned quickly that your church's website is used by your actual members for one of three reasons: to give, to watch a sermon they missed, or to check upcoming events. This means there is an entire part of your website that could be used to reach people who are searching for a place to call home. We started working on our SEO (Search Engine Optimization), we built our website directed toward new people, and we marketed it out through Google ads, social media, and every digital platform you could imagine. Within three months of doing that, our doors started flooding with first-time guests. We went from averaging three to five new people per month to twenty to thirty new people per month.*
>
> *This is the deal. There are hundreds of people every day searching for a place to call home. They are searching for a life-giving church where they can be involved. Whenever we optimize our websites to reach new people, we give that person a chance to get plugged into our ministries and use the gifts God has given them to impact their community. How cool is it that all of that can be accomplished just by optimizing your website to reach new people?*
>
> **—Ryan Keller, Church on a Mission**

Our culture is so used to getting information within a matter of seconds. Whether this is a good thing or not, it's the truth. We are impatient people, and we don't give sites more than five seconds to load or show us the information we are looking for before we move on. The same is true for a church website. If people have to work to figure out where you're located or what time services start, they will become frustrated—uncared for even. And although that isn't the message that you intend to send first-time guests, it's the message they receive. If things as simple as service times and location aren't right there on the front page, do you think guests are going to feel like you're anxiously awaiting their arrival? If the homepage doesn't scream, "You're welcome here! We WANT to have you!" then the implied message, unfortunately, is the opposite.

Every time you share this basic information about your church, it's an opportunity to do outreach, to reach out a digital hand and say, "join us." It doesn't have to be over the top or cheesy; each church has its own "flavor," and that's good. But it does need to be there, loud and clear.

ABOUT US PAGE

When people click on your About Us page, what they really want to know is if they will fit in at your church and align with your beliefs. This page should contain enough information to help them decide that. Consider including your church's mission and vision on this page, maybe a short bio on each staff member, or even some history that is relevant to the way your church functions today. Again, this doesn't have to be lengthy, fancy, or interactive. People want to manage their expectations before they arrive, so give them the most important things to know about your church. And despite the fact that the name of the page is "About Us," remember that this page is serving to answer a first time guest's big question of, "Will I belong here?" Keep them in mind as you create this page.

IMAGERY

Have you ever looked up a hotel online before you stayed there? You may have scrolled through the photos of the rooms, the lobby, the pool, the fitness center, and it all looked incredible! Then you arrived, and you were shocked because the pictures online were not AT ALL an accurate portrayal of reality.

This feels terrible, doesn't it? Yet churches do this all the time with the photos on their websites. Imagery is important! Pictures tell a story, and they help people get a feel for who you are as a church. Do your best to *show* people what a worship service looks and feels like. Make sure your website is an authentic representation of your church community. Free stock photos are a thing of the past, and people can spot inconsistencies between your website and reality. So snap a few pics during, before, and after a service and upload them to your site. Raw, candid photos are what people really want to see!

A website that genuinely reflects your church and is interactive with its traffic has become a must-have. You will be hard-pressed to find anyone who visits your church for the first time who hasn't spent some time on your website. This can be very good news! This is a huge opportunity to make a killer first impression. If optimizing your website is something you're interested in learning more about, check out Boomerang the Course. You can find all the details in the free resource kit. Remember to text "kit" to 816-482-3337 to grab the link!

Adding these critical components to your website will truly help guests to feel obsessed over and wanted! Now, before we move on to Chapter 2, let's do a quick review and look at some key questions your church needs to answer.

THE GUEST-OBSESSED WEBSITE SUMMARY

Your website is the first access point for people who have never visited your church before. In today's world, someone visiting your church without first visiting your website is an anomaly. While there are all kinds of things you can do and say on your website, there are four key things you definitely want to include:

1. Service Times and Location
2. About Us Page
3. Imagery
4. Plan A Visit Feature

Remember that your website doesn't need to be high-tech, custom-made by an expensive web developer, or completely comprehensive. It needs to be a genuine reflection of your church, and it needs to communicate to guests that you want them there.

A key point to remember with your website is that you have *seconds* to get someone's attention before they move on. So please make sure the pertinent information is front and center so people **can't** miss it.

If you're looking for some more guidance with your website, there are a handful of companies that I think are fantastic. If you haven't yet, text "kit" to 816-482-3337 to access the Boomerang Kit where you can see a list of website companies that specialize in church websites.

DISCUSSION QUESTIONS

What is the first thing a potential guest will see when they pull up your website?

Who does your website cater to?

Do you have actual pictures of your church and real people at your church on your website?

Does the language on your website invite people in?

Are your service times and location obvious?

What is the mission of your church? What is the culture? How are these reflected on your website?

Think about your homepage. What needs to be added, and what needs to be removed? Consider asking people who don't go to your church to view your website and give you feedback.

CHAPTER 2

PLAN A VISIT

As our team began to think about a church website as an extension of the church and, therefore, of its outreach efforts, something became abundantly clear. We wanted guests visiting churches' websites! Even more than that, we wanted the number of guests visiting a website to equal the number of guests visiting a church. Boom!

GUESTS ON
OUR WEBSITE

GUESTS IN
OUR CHURCH

Part of that will happen naturally when your website becomes an authentic and genuine representation of your in-person gathering. People want to "try before they buy," so to speak. And a website is an easy way for someone to experience a small piece of your church before taking the big step of showing up.

In the last chapter, we talked about several components, or "must haves," for a guest-obsessed website.

The last one—Plan a Visit—I left out, and I did so intentionally. Not all website features are created equal, and this one deserves its own chapter.

> *The Plan a Visit model has completely changed the way we interact with the majority of our guests. We've been able to connect with, and in some cases pray with, guests before they even set foot in our church. One recent guest in particular saw a video on Facebook inviting his family to plan a visit to an upcoming service at Happy Rock. He planned the visit for his family and attended a service. The following week we continued our Plan a Visit follow-up with his family, and they decided to come to our Welcome Party to find out more about our church. Because of the Plan a Visit model, he was so excited to visit Happy Rock even though he had not been to church in over a decade!*
>
> **—Heather Kemp, Happy Rock Church**

The Plan a Visit feature on your church's website allows a potential guest to make themselves known and for you to connect with them before they even walk through the doors of your church. If your website is the front door to your church, Plan a Visit is the shiny doorknob that lets everybody in.

When Heather from Happy Rock Church added a Plan a Visit feature to her church's website, the church was finally able to reach and connect with people who were thinking about coming but had never shown up. People were expressing their interest by visiting the website, and Heather and her team could get to know them, capture their contact information, start a relationship journey, and invite them to their church.

Plan a Visit fills the gap between people wanting to visit your church and you hosting them incredibly well once they show up. When someone decides they want to come to your church for the first time, actually showing up can sometimes be quite a journey. However, when they fill out this simple form on your website, you're able to pour into them and help ease the transition from thinking about going to church to actually going to church. It's little things like a text message, an email, or a phone call that can make a huge impact!

Let me give you another example of this. As a senior in college, I was looking at all kinds of schools—big ones, small ones, private ones. I had no idea what I wanted. Before doing a college visit, I was asked to register online. Typically, they just wanted to know how many people to expect in their group tour so they could order enough sandwiches for lunch. However, there was one school that went above and beyond. They had a welcome sign with my name on it outside the door to the admissions office. I had my own personal tour guide who showed me around the school based on a survey I filled out that included what major I was considering, sports I was interested in playing, and so on. They bought me lunch (that I got to choose) and had current students who were in the major I was interested in meet with me. I left that visit feeling like that school *wanted* me. I could see myself there, and I felt comfortable with the location and people because I felt like I got to experience so much of it.

I am not championing this kind of "tour guide" approach to church. What I am saying is that when a guest has the opportunity to let you know they're coming, you get the chance to be intentional with them.

I have talked to so many pastors who wish they would have had a chance to speak to this family, that guest, or the one guy who hasn't been to church in years. Imagine what could have happened if a pastor had gotten a notification a few days before that new family was planning to attend on Sunday. What if they had received their name, phone number, and the service they were planning to come to?

One of the pastors we've been privileged to work alongside shared a powerful story with us:

I noticed after church on a Sunday morning, one of our first-time guests was bawling her eyes out. Concerned, I went over and asked if she was okay. She told me she was recently divorced and she and her kids moved here all alone. She said she hadn't felt as loved and seen as she did that day in a long time.

Fast-forward to today. She is now a critical piece of our serve team and thriving in her new community.

THAT's the power of Plan a Visit.

The Plan a Visit model allows your website to interact with the user, to ask them for contact information, or to find out if they have kids who need childcare during the service. It can allow them to pick different times and locations to attend. And the concept can carry over into more than just a guest-care feature. You could use it for VBS registration, for volunteers, for small groups . . . the options are endless. The key is to have multiple access points. We will talk about this in greater detail in chapter 5, Connect Cards chapter, but we have seen the power of this tool on so many church websites.

So how do you implement it? Technically, you can install a Plan a Visit feature on your website in a handful of different ways. However, I think there's really only one good option.

Let me walk you through how Plan a Visit works on the backend through Text In Church's platform. Hopefully, then you can see how beneficial it is for both guests and the church!

 Once you log in as a member of Text In Church, you will see a Groups tab in the left sidebar. Once you click on Groups, you will see a Smart Connect Card button. Clicking this takes you to the card builder. With this builder, you can embed a Smart Connect Card on your website to begin setting up a Plan a Visit feature. (Again, you can use these for so many options other than just Plan a Visit, which we are super stoked about, but we will focus on this feature for now!)

Then, when you open the card builder, it will walk you through four steps:

Step 1 allows you to edit the title of the card and add a description, an option for pre-registration for childcare, and a multiple-choice question.

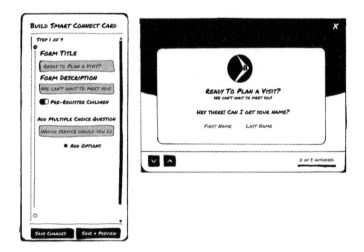

Step 2 allows you to choose where and how you want your Smart Connect Card to be accessed. There are several ways you can deploy the Smart

Connect Card. You can add it in a chat bubble on your website, as a sticky banner that goes at the top or bottom of your website, as a pop-up when someone intends to exit the site, or as its own separate landing page. Each of these features can be enabled or disabled, and you can use more than one at a time.

Step 3 is optional but shouldn't be overlooked! You can do a few different things in this step. You can add a video to the confirmation page (maybe a welcome or promo video your church has created) that's visible after the person fills out the card. You can schedule a follow-up text to be sent after an individual fills out the card. Additionally, you can send a notification to a staff member when someone fills out the card so the staff member can follow up with them. You can decide where you want the confirmation page to be. This can be a page you already have created on your website, or you can use one we have created. Finally, you can add a Facebook tracking pixel to target ads if you so choose.

Step 4 gives you the code to install your newly created Smart Connect Card on your website! You can install the script yourself or send it to the appropriate person who manages your website.

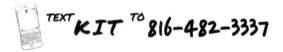

Want to see it in action? Head over to the Boomerang Kit (if you haven't texted in and gotten the link yet, do it now!) and check out our demo site. You can see the Plan a Visit feature deployed four different ways as well as how it can be personalized, etc.

Here are a few of the really fantastic features of the Plan A Visit card:

First, there are many different "access points" on the website. There's the chat bubble, the button, the sticky banner at the top. Some churches choose one, and some use them all. The beautiful thing is that you can play around with the look you like and test what is effective on your site.

Second, when it comes to the actual card, it's interactive. You'll notice it asks for your name first. Then, it uses your name to ask the next question. It responds with "Thanks" or "Great!" It is statistically proven that more people will fill out an information card when they're only prompted for one question at a time. Having a long form on your website that asks for all of this information at once will likely turn someone off, and they won't fill it out.

Third, the information gathered in the Smart Connect Card is organized and can be used later. Building this within the Text In Church system allows you to create a group that corresponds with all your Plan a Visit users. So, anytime someone fills out the Plan a Visit Smart Connect Card on your website, they are automatically stored in your Plan a Visit group in Text In Church where you can schedule emails and texts and start connecting with them before they even show up!

Can you imagine? Right after someone fills out the Plan a Visit card on your website, they get a text that looks something like this: "Hey, Jenna! I'm Tyler, and I work at City Church! We are so excited to have you at City Church this weekend! Keep an eye out for an email from me with some details on parking and kids' check-in. Can't wait to meet you!" This text helps to eliminate the personal vulnerability and fear of attending a new church.

"HEY JENNA! I'M TYLER AND I WORK AT CITY CHURCH; WE ARE SO EXCITED TO HAVE YOU AT CITY CHURCH THIS WEEKEND! KEEP AN EYE OUT FOR AN EMAIL FROM ME WITH SOME DETAILS ON PARKING AND KIDS CHECK-IN. CAN'T WAIT TO MEET YOU!"

If you want to see what to include in an email like the one referenced above, as well as all the other follow-up messages (including ones to people who use the Plan a Visit feature and don't show up!), text "kit" to 816-482-3337 to get the link to your free Boomerang Kit. If you've already gotten the link, you can find these templates under the Resources tab.

The Plan a Visit feature is truly a game changer for

churches. It allows you to engage with first-time guests before they ever walk through the door, which helps eliminate their fears of stepping into an unfamiliar setting. And more importantly, it helps them to feel known and cared for in a personal way. Now, let's review the main points of the chapter and answer some key questions before we move on to our next chapter: The Guest-Obsessed Church.

PLAN A VISIT SUMMARY

When a guest has the opportunity to let you know they're coming, you get the chance to be intentional with them.

Plan a Visit is a way to make your website interactive, and it's effective in engaging your website traffic. We know that seventeen million people who are not churchgoers visit church websites every year. That's a huge opportunity! And this feature allows you to connect with them!

Having Plan a Visit on your website doesn't have to be clunky and complicated. Text In Church's Plan a Visit feature can be built in four easy steps and will then organize all of the information for you.

Following up with the individuals who fill out your Plan a Visit card is key. Whether they show up or not, they've taken the step to fill it out and make themselves known. Now it's your turn to make them feel known.

DISCUSSION QUESTIONS

Do you have a Plan a Visit feature on your website?

Is there anywhere on your site that allows someone to engage with you or reach out?

What do you see as being the biggest challenge of implementing Plan a Visit on your site?

Do you have follow-up messages ready for individuals who Plan a Visit on your website?

Who can head up adding the Plan a Visit feature to your website?

CHAPTER 3

THE GUEST-OBSESSED CHURCH

I was sitting at a conference a couple of years ago at one of the biggest churches in the United States. A member of their staff stood up to give the keynote. She was talking about the dynamics of their church and how, even as big as they are, everyone has an important role to play. She led with the story of her first Sunday at the church. She and her family pulled into the parking lot and were almost immediately greeted by a genuinely kind and warm man. He walked them into the church and connected them to a couple who attended the church regularly. She spoke of the relief she felt to have someone to sit next to, to show her where to drop her kids off and where the restroom was. She also thought about how nice it would be to see a familiar face the next week. As they sat through service, they were enjoying it. Then, the pastor stood up to give the sermon. And who was the head pastor? That same kind, parking lot greeter. She was so impacted by his kindness toward them amidst running such a large church. She said they'd been there ever since.

—Text In Church Staff Member

This doesn't work at every church; I know that. But don't get stuck on the specifics of this example. What I want you to think through is the profound impact being *seen* and *valued* had on this family. This woman plays a huge role in this church now, and that all started because she and her family were thought of before they even walked through the doors. This is a profound example of a guest-obsessed church.

And that is exactly what we will focus on in this chapter: how to be a guest-obsessed church. I can't stress enough how essential it is for your church to have an underlying passion to exist for the unchurched. This is the reason we open our doors each Sunday. To reach the lost and the lonely. To point others to Jesus.

When a first-time guest comes to church, leaving their house should be the hardest part. Once they pull into your parking lot, there should be a trail of breadcrumbs showing them exactly where to go and what to do next. They've done the hard part. They have broken their previous pattern of *not* going to church. They left the spilled breakfast, the piled laundry, the comfort of their own home. They showed up. And once they pull into the parking lot, it's your turn to show up for them.

I'm sure you probably have greeters ready to go every Sunday morning and maybe even coffee and donuts in the foyer. I also hope you have plenty of signage around, both outside and inside, so that people can find parking spots, the main entry, restrooms, the nursery, etc. I know there are plenty of kind, welcoming people at your church who can point guests to all of those places. However, we have found that requiring guests to come forward and ask for help in that way can put them in a vulnerable position.

You might have just rolled your eyes at me there.

Vulnerable? To ask where the bathrooms are? Come on, Tyler.

I know, I hear you. But let me put it this way. Having signage sends the message to new people that this church is expecting them, even hoping for them. It says that this is a place where not everyone has attended their whole lives and can navigate the building with their eyes closed. I understand that every church wants new people, but people who have never been to church or who have been away from the church for any amount of time have real doubts about their place there. Many of the church leaders I work with have found that the more they overcommunicate to their guests, the more significant the impact. Things that we may think are over the top are really appreciated by guests. So, consider all the things/places/people first-time guests come in contact with before the service even begins, and make sure there is someone or something that acknowledges them.

At Text In Church, in an attempt to simplify the prep process for church leaders, we created a "10 Step First Impression Checklist" that makes it easy for churches to go down the list and cover their bases. It includes simple things that make a big impact on how guests perceive your church. Just a reminder, you can download and/or print this resource from the Boomerang Kit. Text "kit" to 816-482-3337 for the link!

10-STEP FIRST IMPRESSION CHECKLIST

1. Website: Updated and Interactive
2. Designated Parking for Guests
3. Exterior Signs
4. Greeters and a Culture of Hospitality
5. Make Check-In Easy
6. Have One Clear Next Step
7. Cue Members to New People
8. Interior Signs
9. Make the Welcome Center Obvious and Accessible
10. Consistent Messaging Throughout Entire Experience: I hope you come, I'm glad you came, and I hope you come back!

Okay, so guests are through the front door, they've parked in their legit, guest-marked parking spots, they've been greeted by outrageously kind people, and they've used the restrooms that are easily accessible thanks to great signs.

What happens next?

This is an important piece. You don't want a first-time guest to have to ask that question. If they are left standing inside your building, wondering, *Where do I go, and what do I do?* you're losing them. There are a couple of ways to avoid this.

We have seen a lot of churches have success with greeters walking first-time guests to their next step. So, if they have kids, the greeter would escort them to the check-in desk for the children's area and introduce them to someone who is working/volunteering there. If they are waiting around before Sunday school, the greeter might walk them to the Pastor of Adult Ministry or a specific class and introduce them to a few people there. If they're there for a service, the greeter finds another member or regular attender and introduces them, so the guests have someone to sit with. This personal connection goes a long way. However, we know there are plenty of churches where this is not feasible. If you can do something like this, do it! But there are other ways to keep communicating to guests without using people.

We encourage pastors we work with to have a designated "Welcome Center" that is obvious and centrally located. This doesn't have to be its own room or anything fancy. It simply needs to be labeled well (yes, another sign) and in a spot where you can't miss it. Some churches have their Welcome Center smack dab in the middle of their entryway. We will get more into the functionality of a Welcome Center later, but this is a great way to have a place for guests where they know they can get their questions answered or where someone will notice them and offer to help.

We also find it helpful to have different ministry areas designated, either somewhere near the front door or via large signage telling people where to go. If a family with elementary and middle school-aged children walk in, they are going to want to get their kids settled first. Make it easy for them! Don't make them ask where they need to go.

I'll say it again. When a first-time guest comes to church, leaving the house should be the hardest part. By taking the time to go through the 10-Step First Impression Checklist, you will take the guesswork and confusion out of the guest experience. Guests will have all the information and resources they need and will know exactly what to do next. And they will feel obsessed over before they ever walk through your doors and long after they leave.

THE GUEST-OBSESSED CHURCH SUMMARY

When someone visits your church for the first time, they are unsure how or if they belong there. It is our great responsibility and privilege to show them all the ways we've prepared for them and hoped for them. It is our job to obsess over them. They need to feel known, noticed, and loved from the moment they pull into your parking lot.

Two of the main ways you can be guest-obsessed at your physical location is with signage and next steps. We've made this easy for you by creating a 10-Step First Impression Checklist:

1. Website: Updated and Interactive
2. Designated Parking for Guests
3. Exterior Signs
4. Greeters and a Culture of Hospitality
5. Make Check-In Easy

6. Have One Clear Next Step
7. Cue Members to New People
8. Interior Signs
9. Make the Welcome Center Obvious and Accessible
10. Consistent Messaging Throughout Entire Experience: I hope you come, I'm glad you came, and I hope you come back!

Having guests in mind as we prepare and produce these things will speak volumes.

Text In Church has an incredible amount of resources on this topic. Make sure you grab your free Boomerang Kit to access these tools and lots of others!

DISCUSSION QUESTIONS

When you think of everything at play on a Sunday morning, are there any disconnects where guests might feel intimidated? Things or places they are unable to access? Are they unsure of what to do next? How can your team work together to create a comprehensive "track" for guests to embark upon?

Do you have signage at your church? Where? Is it clear and easy to see? Is it geared to guests?

What signs are missing? Do you have money for this in your budget? Who in your church could help with this?

What next steps do you want guests to take at your church? How are you communicating that to them? How are they supposed to move forward?

CHAPTER 4

THE WELCOME SPEECH

You never know the state someone is in when they walk through your doors on a Sunday morning. They could be looking for answers and longing for someone—anyone—to see them.

> YOU NEVER KNOW THE STATE SOMEONE IS IN WHEN THEY WALK THROUGH YOUR DOORS ON A SUNDAY MORNING. THEY COULD BE LOOKING FOR ANSWERS AND LONGING FOR SOMEONE—ANYONE—TO SEE THEM.

When we acknowledge guests and sincerely tell them, "we are so glad you are here," this could be life changing for them. That's why we believe setting aside some time and intentionality for a warm welcome at the beginning of your service is important. It could make an impact on your first-time guests in a meaningful and memorable way.

Why should this happen at the start of the service? Because the beginning of the service is a critical time to connect with guests. We know that people don't like to feel targeted. The days of "if it's your first time here, please raise your hand so we can welcome you" are long gone. But a Welcome Speech is a way to make sure first-time guests know that they are seen and celebrated.

> We were new to the area and so lonely. I had just cried to my husband that morning that I missed my community so badly and was worried we might never find another like it. We had attended plenty of churches, but we just felt invisible—in and out without any interactions. We tried a new church one Sunday, and a couple stood up to welcome everyone to the service. The woman spoke exactly what I had been feeling: she had once felt isolated, lonely, and longing for community. They had found it here and were so hopeful that they could pour it out to any new people in the sanctuary. I sat there with tears rolling down my cheeks. I felt like I could hear God whispering, "I see you." This couple had never met us before, but they had walked the same road and found their people at this church. I was filled with hope and excitement. We ended up attending that church for years and became dear friends with that couple.
>
> **—Anonymous Visitor**

Okay, so, let's recap what we've done so far. Your guests have now dropped their kids off and are ready to head to the service. They knew exactly where to park because of your excellent parking lot signs; they were greeted outside the church by a super friendly family who walked them to the childcare wing; they got their kids all checked in thanks to the help of a childcare volunteer; and they found that sweet greeter family waiting for them outside of the sanctuary. They're now plopped down next to some semi-familiar faces ready to see what the service is all about. (Of course, this example is going to look different based on the needs and abilities of your church, but you get the gist.)

Enter: The Welcome Speech. We have developed what we call the Cornerstone Method to The Welcome Speech because it's foundational for all the other follow-up that needs to happen. The Welcome Speech has a lot packed into it, so I'm going to break it down into three distinct parts:

1. Welcome
2. Connect
3. Desire and Why

WELCOME

Sounds simple enough, right? And it is. Nothing earth-shattering here. The person delivering The Welcome Speech should stand at the front of the congregation and give a warm and genuine welcome.

As an aside, the jury is out on *who* should be delivering The Welcome Speech. Some churches rotate their staff, some use members of the church, some use entire families, some have the lead pastor do it every time. I like the relatability of having members, both new and old, give The Welcome Speech. However, I know that there's value in having just staff do it. That's something that you can figure out at your specific church.

Whoever does it should give a special shout-out to your visitors. Maybe the first part sounds something like this:

> *Good morning and welcome to City Church! My name is Tyler Smith. I've been a part of the City Church family for seven years now. If you're new here, we want to give you a special welcome. We are so grateful you decided to worship with us this morning.*

This is impactful, I promise you. This is a time where you can implicitly say to your guests, "We see you. You belong here." No matter how tough or apathetic someone may appear, this is the heart's desire of everyone. We have to remember our first-time guests aren't just people who are new to the area and looking for a new church home. The first-time guest at your service could be visiting church *literally* for the first time in their life. That's super exciting! We want them to know that they're welcome here.

CONNECT

The next part of The Welcome Speech is meaty. You need to connect with first-time guests a bit. By letting them know that you're human too, it will set them at ease. You can share a bit about why you love your church and how you are involved. I also think this is a great time to address some of the unspoken "rules" or traditions at church that new people can be really intimidated by or just plain weirded out by. Here's an example of what connecting with first-time guests could look like in a Welcome Speech:

As I mentioned, my family and I have attended City Church for seven years. My wife and I get to serve as life group leaders as well as in the children's ministry. We have loved how focused City Church is on serving our community. You can find all of our local service opportunities on the back of the bulletin. Again, if you're new, I want to thank you for being here and give you a little bit of an idea of what this hour will look like. We like to start the service off with singing. People worship in a lot of different ways, so please don't feel compelled to do anything specific. Then, we pray. This is a time when you can interact with the living God—however you best do that. We also have a Scripture reading. At that time, we all stand in honor of hearing God's Word. And then Pastor Jake will bring us a message. Don't worry. We will tell you before stuff happens. Basically, just know that you can't mess this up.

This may feel unnecessary and wordy, but it can help alleviate a lot of nerves. For those of us who grew up in church, we can take for granted how uncomfortable a traditional church service can be. It doesn't do any harm to unpack it a little bit so people feel more at ease and know what to expect.

DESIRE AND WHY

The third piece of the Cornerstone Method is where you tell the guest what you want from them.

This is also where I introduce the next crucial part of the Gather step—the Connect Card. We have an entire chapter dedicated to the Connect Card, but stick with me while I talk briefly about it here.

One of the worst things that can happen after someone has attended your church for the first time is for them to be unsure of how to move forward. Let's say they loved the sermon, felt connected to God through the music, and met some great people. But then they weren't sure what to do next. If there's no next step, we are doing them a huge disservice. We are leaving it in their hands to find the motivation to keep coming back, to keep being intentional, to keep pushing themselves into belonging there. We have to be very clear about what we want them to do and why they should do it. This is a time to metaphorically open your arms and embrace them! Here's what that might look like in The Welcome Speech:

If you're new here, we would love the opportunity to connect with you. In the seatback in front of you, you will find a "Welcome Card." Please jot down your name, number, and email on that card and drop it in the offering plate when it comes around. We'd love the chance to follow up, share some info about the church, and answer any questions you may have. Or if you have your phone with you, you can text "NEW" to 816-482-3337. This will also provide us a way to follow up with you. Finally, please drop by our Welcome Center after the service. If you go out the doors and straight back, you will see a big table with a Welcome Center sign over it. You can't miss it! A few of our pastors will be hanging out there after the service and would love to meet you and give you a little gift.

And there you have it. In just one to two minutes at the beginning of your service, you can create a huge impact on first-time guests. This short Welcome Speech, as well as everything else we've talked about so far in the Gather step, works to help build trust with your guests and make them feel valued. It will also increase the likelihood that guests leave their contact information, and that is the key—the cornerstone if you will—to connecting with them.

THE WELCOME SPEECH SUMMARY

A Welcome Speech is the perfect opportunity to recognize your guests without making them come forward in any way. A public, individual recognition can be incredibly embarrassing and isolating to some guests, but a Welcome Speech is a great way to tell them that you see them and you are thankful they are there.

The three keys to crafting a perfect Welcome Speech are:

1. Welcome
2. Connect
3. Desire and Why

All of these pieces work together and build upon the foundation you've laid on your website, communicating that your church is excited and prepared for new people!

DISCUSSION QUESTIONS

Does your church give a Welcome Speech every Sunday? What does it include? Who are you speaking to in your Welcome Speech?

Is the language used understood by people outside of your church, or is it full of Christianese and insider lingo?

If you don't do a Welcome Speech now, how can you carve out a minute or two in your service to add this in?

How could you cater the three sections of The Welcome Speech to your church? What would be the most important things to include?

1. Welcome

2. Connect

3. Desire and Why

Who at your church is best equipped to oversee or execute this week in and week out?

CHAPTER 5

CONNECT CARDS

One hundred percent of the time, we will fail to adequately follow-up with a first-time guest if we don't have a tool for gathering their contact information.

> WE, 100% OF THE TIME, WILL FAIL TO ADEQUATELY FOLLOW-UP WITH A FIRST-TIME GUEST IF WE DON'T HAVE A TOOL FOR GATHERING THEIR CONTACT INFORMATION.

This seems obvious, but how many times have you met someone on a Sunday morning and panicked because you didn't have your phone to capture their information? So, you repeated their name over and over in your mind to try to remember it so you could look them up on Facebook and connect that way. But you can't find them on Facebook, so you ask the Children's Director if there were any new kids with the last name Smith. Or maybe it was Smitty? No, it was for sure Schmidt.

You get the point.

It doesn't have to be this hard! Connect Cards have been used in the church space for . . . forever. However, we've learned that many people will not take a pen and fill out a piece of paper. I'm one of those people, so I get it.

The solution? Expand the idea of a paper Connect Card into different formats so techy guys like me never have to use a pen and traditional people like my wife can still fill out a paper one with her perfect cursive. People are wired differently, so give them options.

A Connect Card is simply that, a point of connection. We are not trying to populate our database, determine how many kids guests have, or anything else. We just want our First Impressions/Membership Team to be able to reach out one on one. That initial conversation will reveal specifics. If guests have kids, then great; we'll put them in touch with our Children's Ministry so they can learn more about that. If they are new to the area and looking for a church home, our membership coordinator can reach out. The point is, we are not trying to capture all their information on a Connect Card. We simply want to know the best way for a person to reach them during the week.

Take a look at a few of our church's Connect Cards. (The original images can be found in the Boomerang Kit!)

GOOD   BETTER   BEST

The two on the left are older and the one on the right is more recent. Note the differences you see.

The point of showing these is to illustrate just how easily we can fall into the "that's the way it's always been done" trap. The twenty-year-old Connect Card at the far left is an excellent example of that. The only significant difference between that one and the previous version (from 1999) is that the card didn't have a blank for email or cell phone!

During the welcome time of our service, everyone is encouraged to fill out the Connect Cards if they have a need, a prayer request, or an interest in serving, becoming a member, or baptism. First-time guests can drop the card off at the info desk in the lobby area where they will receive a free gift and more information about the church. Our gift is a ceramic coffee mug with the church logo, a business card with the contact list for our ministries and service times, and a small packet of information about our ministries, community partners, and church membership.

During the week, our membership coordinator compiles the collected cards, makes contact with the individuals and families, and passes their needs and requests along to the ministries that can best serve them.

—Richie Huval, Harmon Hill

In this chapter, I'm going to tell you about three different formats for the Connect Card, but remember, they're all working to accomplish the same goal—to gather information. I don't believe one is better or worse than the other, but I do believe in providing several options for first-time guests.

Also, one other note of caution before I dive into the three formats. Don't ask for too much information. Remember that these are your first-time guests. We don't need their physical address, all their children's names, their church history, or their blood type—okay, hopefully none of you ask for that, but you get my point! You wouldn't propose marriage on the first date with someone. So pump the breaks, play it cool, and get enough information to stay in contact with them. I recommend gathering their name, phone number (cell if you can), and email address.

Now let's dive into three effective formats for Connect Cards.

PAPER CONNECT CARD

This is old faithful right here. If you've been in the church space for any amount of time, you likely know what a paper Connect Card is. These can be placed in the seatbacks, in or on the bulletins, part of check-in if guests drop children off, or at the Welcome Center. The important thing about paper Connect Cards is you have to have a good method for collecting them. If you want to have paper Connect Cards in as many places as possible, make sure you still have a designated spot for them so staff/volunteers can pick them up and track that information. You can call these whatever you want—a "Welcome Card," "Get to Know You Card," or "Connect Card"—and you can brand them to match your church. There are great templates for these as well as plenty of free platforms where you can design your own. (I like Canva!)

SMART CONNECT CARD

The Smart Connect Card is a newer concept that churches absolutely love! It allows churches to offer their Connect Card on a device (tablet, phone, website, etc.). This is huge for outreach events! Let's say your church does a big Fall Festival. With a Smart Connect Card, people who come from the community can check in to the event with an iPad you have set up next to the apple cider table, and then you have the ability to call or email them and invite them to church! Pastors also like using the Smart Connect Card on devices at their Welcome Center. This allows guests to meet real people from the church, making the exchange of information feel less unnatural.

What's especially fantastic about the Smart Connect Card is its "success rate." Traditional digital Connect Cards looked more like this, with all of the fields on one page:

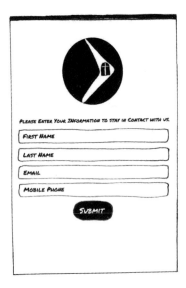

The Smart Connect Card only asks one question per page, which creates a statistically higher likelihood that people will finish filling it out. A 2019 article published by VentureHarbour looked at several different case studies and found examples where multi-step forms increased conversions by 35 percent for BrokerNotes, 59 percent for Vendio, and 214 percent for an Astroturf company.[1] So why is this the case with multi-step forms? Here are a few possible reasons:

- Multi-step forms reduce psychological friction.
- The first impression appears less overwhelming.

---

[1]   Aaron Brooks, "5 Studies on How Form Length Impacts Conversion Rates," VentureHarbour, March 13, 2019, https://www.ventureharbour.com/how-form-length-impacts-conversion-rates/.

- The progress bars encourage users to complete the form.

With a traditional Connect Card, people see all of the questions at once and decide not to fill it out. The Smart Connect Card, however, is more user friendly. There is personalized text prompting guests for information, so it feels more interactive. Let's take a look.

This personal and unintimidating approach to the Connect Card truly is smart!

TEXT TO CONNECT

Text to Connect is the cat's meow if you ask me. It is what I alluded to in The Welcome Speech. But instead of explaining it, why don't I just show you?

Grab your cell phone.

Text "Boomerang" to 816-482-3337.

Did you get a text response? The link you received sends you to our Smart Connect Card. This is exactly what so many churches are doing on Sunday mornings—gathering guests' information via text. People can text in and fill out the information on their phones, right from their seats! The churches we work with then have a system that collects all the information gathered by the Smart Connect Cards and immediately plugs those people in for six weeks' worth of follow-up. Thousands of churches across the country are seeing massive success with this.

TELL THE WHY

Something I do hear from pastors all the time about Smart Connect Cards is "people just won't fill them out." Maybe they text in, but they don't take the next step to fill out the card. Perhaps they put their name on a paper Connect

Card and leave it but don't leave any follow-up information. There will be those instances, and I know it's frustrating. For some people, it just takes time for them to trust you enough to leave their information. However, usually my conversations with church leaders go something like this:

"Tyler, I'm not getting any information from my first-time guests."

"Ugh, so sorry. I know that's frustrating. Do you have the Connect Cards offered in multiple formats?"

"Yep, we do."

"Do you introduce them in your Welcome Speech?"

"Uh-huh."

"Do you tell people why you want their information?"

"Uh . . . no."

This is a common problem! It is a huge ask of someone to give their phone number when they've just come to your church for the very first time. They don't know why you want their phone number. So, as we talked about in The Welcome Speech chapter, you must give them some context around why you want information from them. If they've never been part of a church, they don't know how this works! Eliminate the unknown. Left to our imaginations, we almost always think of the worst-case scenario. Don't let your guests wonder *why*. Tell them! This will help you build trust with first-time guests.

As we wrap up the Gather section of this book, think back on all we have discussed so far:

- The Guest-Obsessed Website
- Plan a Visit
- The Guest-Obsessed Church
- The Welcome Speech
- Connect Cards

You now have all the tools you need to gather information from first-time guests. Spend some time with your team reflecting on these strategies and discuss how you can implement them effectively at your church. This will lay the foundation for the Connect section we'll dig into next and the Build section we will get to later.

CONNECT CARDS SUMMARY

I have seen incredible success with the use of the three different Connect Card formats:

Paper Connect Cards: the tried-and-true paper version that has been around for generations. Be intentional about where and how you use these so you don't physically lose them.

Smart Connect Cards: a digital version of your paper Connect Card. It's interactive, personalized, and syncs directly into your follow-up system.

Text to Connect: a "text in" phone option that uses a keyword and gives access to the Smart Connect Card. It can also be used for guests, events, volunteering, etc.

The three most important things to remember about Connect Cards are:

1. Use more than one kind. People like what they like, and no two people are going to like the same thing. Have options so you don't lose anyone.
2. Don't ask for too much. To effectively follow up with people, we really only need their name, phone number, and email address. Keep the barrier to entry low—you can always ask for more info later.
3. Tell people *why*. It's crucial that people understand why you want their information. This will help build trust and eliminate fears for first-time guests.

DISCUSSION QUESTIONS

Are you currently using Connect Cards?

Do you have a process in place for collecting the cards and organizing the information so you have everything you need to follow up with your first-time guests?

Are there any holes in your gathering process?

What feedback have you gotten on how you are gathering information and organizing it to follow up? Is it working for your staff? Is it working for guests?

Who is in charge of this process? Do they need help or additional resources?

How many guests, on average, do you see on a Sunday morning? Of those, how many are leaving their information? If this number is lower than you'd like, how can you improve your process?

CONNECT

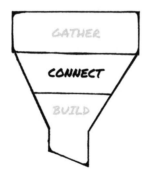

My husband and I were visiting a church for the first time, having just moved from out of state. We have two little girls, and we were blown away by the children's ministry. We were noticed and greeted right away. A woman signed us in and walked us to each of the girls' Sunday school rooms, introducing us to the teachers, giving us a tour, and making sure we had all our questions answered. After the service, we were excited to go pick up our girls! They both had such a fun time, and they both had a balloon tied to a little gift bag. We couldn't believe the thoughtfulness and generosity. The bag was filled with superhero stickers, follow-up material for us to talk with the girls about the lesson, and a new water bottle. The girls ran around the foyer with their red balloons like it was Christmas morning! We couldn't help but notice how many people came up and introduced themselves to us. They asked if we were new and how we liked it; they invited us to come back; and they were so genuine. Upon returning the next Sunday, I told one of the pastors I was so impressed by how friendly everyone was. It's not a small church, so I was amazed people could pick out guests like that!

The pastor smiled and said, "well, the bright red balloons are a good giveaway."

—Anonymous Visitor

love that story! Something as simple as a balloon triggered a churchwide effort to obsess over first-time guests, making sure they felt noticed and welcomed. This is a perfect segue into our next section—Connect.

Connection can and should happen all over your church—during service, before and after, throughout the week, etc. This is a key part of *Boomerang*. We in no way want to replace connections that happen organically in your community. However, there are a few strategies we've seen that make a huge difference in a church's ability to connect with those who otherwise might slip through the cracks. And we've seen technology used as a tool to facilitate *more* connection. So that's what we are going to dive into here.

Plenty of people who will visit your church know what they want; they will even attempt to make connections their first time visiting. They will ask questions and seek out information, and, even in spite of oversights, they will end up becoming a part of your community.

However, there are also plenty of people who have doubts at each and every point. You know who I mean. You can probably picture some of their faces right now. They are the people you always wonder about. They're the ones who walk out the door and your stomach drops because you never got the chance to meet them . . . and you're not sure you ever will. They're the ones who promised their spouse they'd give church just one more try. Or the ones who came only because their grandma begged them to come. They're the ones who are going through horrible life circumstances and are desperate for hope, so they showed up to see what this church thing is all about. The truth is that church attendance is declining rapidly. And if we aren't in active pursuit of people who have fallen away from church, then the world is. We need to help people understand that we want them at church because we care deeply about them.

So, all of these strategies aren't ways to make your church hipper or to make you rethink for the sake of being relevant. They're actual proven ways to help people who feel disconnected from church get drawn in again.

And that's the whole point!

CHAPTER 6

WELCOME CENTER

The first major, strategic space I see for connection before and after the service is a Welcome Center. The Welcome Center, as I mentioned in Chapter 3, should be a one-stop shop for your first-time guests. Naming this something similar to "Welcome Center" communicates that it's a spot for newbies without totally spotlighting them. And it doesn't have to be its own room! It can be a little café area, a room, or a table in the middle of your foyer. The details are up to you! The Welcome Center is a great spot to send people after the service with some sort of an action (Remember what we talked about in The Welcome Speech?).

Here are a few ideas:

"Stop by the Welcome Center so we can meet you and give you a small thank-you gift for being here!"

"If you're new here and interested in meeting some great people, they're waiting at the Welcome Center and would love to have you join them for lunch!"

"Please feel free to stop by the Welcome Center after service for some coffee and donuts; we'd love the chance to connect with you!"

The options are endless here; you just need to give them a reason to go to the Welcome Center. The reason acts as an icebreaker. Instead of a guest awkwardly hanging out there and hoping someone notices them, they can walk there, knowing someone's expecting them. Shoot, they can even grab the gift and leave if they don't want to talk to anybody! Again, it's eliminating the question: "what's next?"

In addition to the Welcome Center being a great buffer to connect first-time guests with a staff or church member, it is also a great place to have information about ways people can get plugged into your church (ministries, outreach, small groups, etc.). For example, you can have signups for Sunday school or your church's upcoming Vacation Bible School. Maybe you do a local service project every month or have a Next Steps class you want everyone to take. This is a great spot to let people know what's going on in your church!

One thing I would highly recommend having at your Welcome Center is your Connect Cards. Whether it's a

Smart Connect Card on a tablet, paper cards on the table, or a combination of the two, this could be your chance to make a face-to-face connection with someone who hasn't been willing to give their information yet. Maybe after meeting a kind person or two, they will feel more comfortable receiving some follow-up. The churches we work with are asking people whenever they come to the Welcome Center if they feel comfortable leaving some contact information. They will either say they "already left information" (via a paper Connect Card or Text to Connect), "no thanks," or "yes," grateful you want to stay in touch with them! Sometimes I think this step feels hard for some churches. But there's no risk, and the opportunity for reward is huge!

The most success I've seen at a Welcome Center is when it's in a central location. It makes a ton of sense for a guest to walk out of service and inevitably find themselves at the Welcome Center. If it's centrally located, there is also more traffic in general, creating a more natural gathering place for members and guests alike. We have seen a lot of the organic connection happen here simply because people all file out of the sanctuary and land in this same spot.

I think an important point to make here is that your Welcome Center is pretty useless without people at it. Don't send first-time guests to a Welcome Center empty of staff or volunteers. A considerable part of guest care is the people who are hanging out, watching, and waiting to love on the guests who come. Think about how you're training your guest-care volunteers, if you have any. That is an essential piece to the Welcome Center—obsessing over how you care, communicate, and connect with first-time guests.

The Blue Room is a one-stop shop for anyone in the church. He says, "If you're a guest, you head to the Blue Room. If you want to join a life group, you head to the Blue Room. It's also a space for things like a guest speaker's book sales and any other one-off need. It keeps us from setting up tables in the lobby. No matter if it's your first time or you're an active member, everything you need is in the Blue Room." (Check out the actual images in the Boomerang Kit!)

—David Urzi, Cape First Church

WELCOME CENTER SUMMARY

A Welcome Center, along with well-trained staff or volunteers, creates an excellent spot for guests to come and meet new faces, leave their contact information, and find out their next step. And there are many different forms this can take:

- A separate, more private room or space
- A table in an open space
- A general area where you aim to gather after service

- A multifunctional space (maybe it's your fellowship hall that doubles as the Welcome Center before and after services)

My biggest recommendation is that it's easy to see and find. Guests are less likely to travel down the stairs, walk through a hallway, or make six left turns. Make it easy for them!

DISCUSSION QUESTIONS

Where do you aim to connect with guests now?

What's the next step you want guests to take after service? How do you communicate this to them?

Who at your church is exceptional at relationship-building, connection, and inclusion?

What does guest care look like at your church on a
Sunday morning?

How can you utilize a space you already have in your
church for connecting with guests before and after
services?

CHAPTER 7

NEXT STEPS

I mentioned something in The Guest-Obsessed Church chapter that I think is critical in informing how you transition first-time guests into, hopefully, regular and active attendees.

You need very clear, accessible next steps.

I'm going to dissect the above statement, so you understand what I mean and how important this is. The reason your next steps need to be explicit is twofold: to eliminate confusion and to maximize connection opportunities. If a guest comes to your church for the first time and visits your Welcome Center, you need to resist the temptation to give them information on every program your church offers. A first-time guest, in most cases, is nowhere near ready to get plugged in, serve, or learn where their financial contributions are going. They want to meet people. They want to know what this church is about and figure out if they fit in. The more options and opportunities you give them, the more overwhelmed and confused they will become.

Therefore, you should have one next step you want all first-time guests to take. This can be as easy as "fill out the Smart Connect Card" or "stop by the Welcome Center." The strategy, then, is to have a next step ready after this first one. Looking at the example from Summit Park, they have all guests stop by the guest tent, and then they invite guests to attend the Welcome Party. There is no talk of small groups or serving or any other area we want people to plug into until step 3 or 4. Then each step after takes everyone through the same process, ensuring they are connecting with key people at your church as well as other guests.

Now, let's talk about making the next steps accessible. You want to remove as many barriers as possible when creating next steps. Think about things like childcare, food, location, etc. We know you can't make every event work for everyone, but you can be conscious of the demographics at your church and what their greatest needs are.

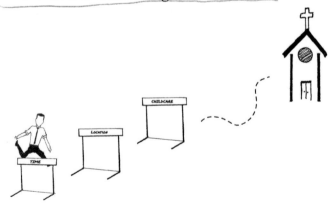

All first-time visitors are invited to go to a guest tent after the service. There, we have a gift for them and friendly volunteers to connect with them. We also tell them, "stop by next week, and we have a T-shirt for you!" So, there's always this connection point for guests. Then, we invite every visitor to our Welcome Party. The Welcome Party has now taken the place of a 101 class. We wanted more people to show up, so we started to think through barriers and what was keeping people from taking a next step. Honestly, we thought people would be more excited to come to a party than a class. So that's what we did, and people love it! We host the Welcome Party once a month. We cater in BBQ, provide childcare, plus, there's lots of dessert and games; it's a lot of fun. Pastor Scott shares just a little bit at the end of the night. He tries to hit on who Summit Park is and what we are about, and then we invite people to take their next step.

This has been really successful for us. People love it. I think it works for us, too, because it fits our brand as a church. We are more of a youth group for adults, geared more for people who are new to church. We talk about it like it's a date night. We really want it to feel like it's a fun party, date-night type event. Typically, there are round tables set up for eating and games, and there is an assigned table host—someone who is in leadership at the church. This provides another personal connection, a face and name and phone number that they can connect with. We do guest follow-up via text and email and all that, but it's nice to have guests leaving with an actual person they connected with and who they can reach out to again.

After the Welcome Party, the next step is a Connect Night, which is like 201. We felt like it was overwhelming to ask for a commitment to small groups or volunteering at the first event, so the Connect Night is when we try and plug people in, in an intentional way. We hold these once a month as well.

—Summit Park Church

* Great Format *

Are you leading an older church? If so, maybe you have a Sunday afternoon gathering. Are you leading a church with lots of young families? Consider providing childcare so parents can relax and actually connect with other people. Are you leading a church with lots of singles? Maybe have your event at a trendy coffee shop close by. Be creative and intentional. These activities don't have to be expensive and extravagant. In fact, they shouldn't be, because they need to be offered on an ongoing basis.

And just a quick aside on this point: this is a great way to get active members/attendees involved. You can organize a team of volunteers who plan, coordinate food, attend and connect with guests, and so on. Guests love meeting the church's leadership, but they also love meeting other "regular" people in the church.

When first-time guests don't have accessible next steps and clear ways to connect, they struggle with knowing how to plug in. Let me tell you a quick story of a time someone on my staff struggled with this very problem.

One of my staff members grew up in church. She went to a Christian college, volunteered at church as an adult, and now works for a company that serves churches exclusively. So I think we can rate her comfort level with churches pretty high, right? She and her family recently moved to another state and, upon visiting a church for the first time, she found herself overwhelmed by how to get her family plugged into this community. There were classes that met on Sunday mornings, community groups that met at people's homes, and other small groups that met at the church at random times throughout the week. She was excited about a couple of different groups but was always directed to the website to "sign up." She had no opportunity to meet people before signing up for their group, which felt like a big jump—not to mention, risky and vulnerable. These are not things you want guests to feel at your church!

We are talking about someone who is committed to their tradition of attending church. And, still, she was at a loss for how to immerse herself into this new community. Connection is hard and intimidating, and even though we crave it, we aren't always the smoothest creatures at creating authentic community. Expecting people to meet organically before and after service is unfair. Many churches with multiple services have to move people out of the building to allow the next service to come in, so meeting people who might attend a certain small group or Bible study can be a real challenge.

This woman was never going to walk away from church because of her lack of next steps, but someone who had nothing keeping them there would. When the church plans these activities, specifically for new people, you eliminate the question: "Where do I belong?" Sometimes the first step to join a small group or attend Sunday school—or however your church structures your adult learning and community building—can feel intimidating. It's an established group, so they'll be the obvious new person who has to work to fit in. But when you have these baby steps, these casual things specifically designed for new people, there's a sense of relief knowing everyone's in the same boat. And then it's a great spot for organic connections to start happening and for people to keep coming back. Boom!

NEXT STEPS SUMMARY

You need very clear, accessible next steps. This helps eliminate confusion and maximize connection opportunities.

Resist the urge to give first-time guests too much information. If you give too much information, they will feel overwhelmed.

Make next steps accessible by removing barriers. This means you will want to consider childcare, food, and location when organizing activities. And these activities don't have to be expensive or extravagant, as long as they're created with new people in mind.

When first-time guests have clear and accessible next steps, they will know exactly how they can plug in and won't have to ask, "Where do I belong?"

DISCUSSION QUESTIONS

Do you have clear and accessible next steps for first-time guests? If so, what are they?

1. Connection Center to get a gift
   — After that what's next

Do you provide too much information to first-time guests, potentially overwhelming them?

What do you currently do to help first-time guests get engaged and connected at your church?

Where/how do they access these opportunities? (online, sign up in person, show up, etc.)

What feedback have you gotten from members and guests about their transition to your church?

CHAPTER 8

A FOUNDATION FOR GUEST FOLLOW-UP

As we dive into follow-up strategies for first-time guests, take a minute to think about what your church is already doing. How are you currently trying to connect with first-time guests, and what is your system for getting them to return? When you have people who are coming to your church for the first time, many of them have spent an entire lifetime *not* going to church. Others have just been out of the routine for a few months. Either way, they have started to fill that time slot with something else. Even if it's just a morning where they sleep in or spend time together as a family, they have to make a lifestyle change to start a new habit of attending church.

And unfortunately, the statistics of first-time guests who return a second Sunday are staggeringly low. Depending on what you read, only 10 to 16 percent of first-time guests return to church. I don't have to tell you the impact this has on church growth and the health of local churches. If we don't have new people coming to church and staying at church, our churches will die out. And why is this? Well, I don't believe the message of the Gospel has gotten any less powerful. And I don't believe that churches put on "bad" services. I don't think Christians are bad people, and I definitely don't think people are against being involved in a community.

So what do I think the answer is?

I think our world is louder than ever.

Information has never been as accessible and constant as it is right now. People have the opportunity to do and buy and connect anytime, from anywhere. And churches, frankly, haven't kept up. A good friend of mine in the church communications space says that churches are perfectly designed to reach the world as it was fifty years ago.

That resonates with me! You see, I walked away from church, too, for a time. I had my gym calling me, making sure I'd be back for another class. I had my professor email-ing me about his online office hours and letting me know he was available if I had anything I wanted to discuss. I had my local park and recreation office texting me about the seasonal teams and asking if I wanted to sign up. What did I hear from my church?

Nothing.

I know sometimes as church leaders we err on the side of passive because we don't want to be pushy. We want people's own faith journey and personal convictions to bring them back to church. But that message can also be received as apathy—or worse, neglect. I felt like I didn't mean anything to my church if I wasn't involved and pres-ent every Sunday like a "good Christian."

Now, it's true that some people are going to want their space. They won't want to hear from you multiple times, but isn't it better to allow them the opportunity to opt out of conversation than to risk leaving them feeling unwanted and forgotten? I, and the pastors I work with, haven't found too many people upset by a church's attempt to reach out. I think it's interesting that we worry about seeming inauthentic or salesy when we aren't selling any-thing and we do authentically want these people to come back. Right? We are *inviting* them to something; who doesn't appreciate that?

And that's the key word here: inviting. Please note I didn't say *invite*. I also didn't say *an invitation*. I said *invit-ing*, suggesting an ongoing process, a continual ask.

95

We were nervous to start Text In Church because of the fear of "harassing" people with so many texts and emails over the six-week period, but we were wrong! Here are some of the comments from people about us contacting them for six weeks:

"I have never felt so loved."

"I have never had a Pastor and his wife reach out to me personally, and so I thought, I have to go back to that church."

"You guys do a good job of reaching out."

"I love getting those texts from Holly. I wish they wouldn't stop. It's nice to be remembered."

"Thank you for reaching out."

We have new guests who now text us and tell us why they can't make it on a Sunday. I think that points to them having feelings of accountability. In a day and age where church attendance is declining, they want us to know why they are missing and that they'll be back the next week. That's a win in our book!

—Holly Howard, Bridge of Hope Church

Why is this so important?

Well, when you consider a first-time guest, meaning someone who's never been to your church before, you've got to believe that they're at your church for a *reason*. This goes back to the 5-D Theory I mentioned in the Introduction. People are most likely visiting your church because of a divorce, a death in the family, a displacement (a move), a development (maybe a child or marriage), or a disaster. There has been a shift in their life—usually a dramatic one—and that has jarred them into getting up on a Sunday morning and walking through the doors of a church. So, while there has been a significant change in their life, life hasn't necessarily slowed down to free up an extra couple of hours in their week. The competition for time has never been fiercer. So even people who grew up going to church find themselves missing service nearly as often as they attend. In fact, the national average for church attendance for members is only about 60 percent. The point I'm trying to make is this: people are showing up at your church equal parts desperate and distracted. They are longing for an answer, a new way to deal, or just some peace, while they are trying to balance a full plate, a busy schedule, and a constant to-do list.

Do you see where I'm going with this?

We can't do anything about the reason they showed up at church. We can't fix the pain; that's God's work. But we can do something about their detached, "oops, we missed church again" mind-set.

What we've discovered is a proven follow-up *plan*. And I'm going to lay the entire thing out for you, step by step, text message by email. Are you ready?

A FOUNDATION FOR GUEST
FOLLOW-UP SUMMARY

Only 10 to 16 percent of first-time guests return to church.

Our world is louder than ever, and information has never been more accessible or constant. But churches haven't kept up.

Following up with people after they visit for the first time isn't pushy! If we don't follow up with them, they may think we don't care. This can be perceived as apathy or neglect.

This is why we must continue inviting them with an ongoing process.

DISCUSSION QUESTIONS

What is your church's current follow-up plan for first-time guests?

What about it is working?

What about it isn't working?

Do you get feedback from first-time guests? If so, is it mostly positive or mostly negative?

CHAPTER 9

A FOLLOW-UP PLAN
THAT WORKS

I was visiting a church with my younger brother. He had walked away from his faith years before and, although he had made peace with God, he was just taking baby steps to reconnect with a faith community. I watched him during the service and could tell he was impacted. I remember praying that this would be the church, this would be the one that brought him back. He told me after that the sermon spoke directly to his heart. He hadn't heard a pastor speak without feeling condemned and judged in so long. He felt like God was alive in that room. He spoke with an associate pastor after the service, was invited to a small group that week, everything! I was thrilled. I remember talking to him the next week and asking how church was.

"Oh, I didn't make it," he replied.

"What? I thought you loved it?" I asked.

"I did. I do!" He said. "I was out late Saturday night and just needed some sleep. I'll shoot for next week."

"Did you try that small group?" I pressed.

> *"No, I forgot the guy's name and I lost my brochure, so I wasn't even sure where to go or anything."*
>
> *I was heartbroken. I'd heard all this before. He never went back to that church. And my worry that he won't find a church home continues. The impactful sermon wasn't enough. He had heard nothing and, therefore, did nothing.*
>
> **—Text In Church Staff Member**

I wish I could say that this is the first story of its kind. But I've heard of this situation countless times, as I'm sure you have too. Churches have worked so hard on their Sunday morning services. Guests love going to a church where people are friendly. It's great to shake hands and learn a few new names. But then 144 hours later—when the kids are tired and whiny and don't want to go, and grandma called and wants to swing by for coffee, and mom's favorite class at the gym is at noon—what's going to win out? Will it be that one couple we met six days ago who was really nice or the routine we already have of *not* showing up. We have to insert ourselves into that routine a bit. We have to become part of the day in, day out noise, otherwise, we get totally drowned out!

So this follow-up plan consists of the right kind of messaging, delivered in the right frequency, over the right amount of time.

Think of it like a formula: content + frequency + duration = YAHTZEE. Or, better yet, BOOMERANG!

As far as content goes, the key is to diversify. Obviously, if you know anything about what I do for work, you know I think texting is a great idea. Twenty-three billion text messages are being sent each and every day. Text messages have a 98 percent open rate; email comes in second at closer to 20 percent! Texts are read within the first three minutes of being received. I mean, do I really need to go on?

But I will . . .

If you were to try to connect with a friend over lunch, how would you set that up?

Maybe a handful of you thought, *I'd call them.* But I'm going to guess that almost all of you thought, *I'd shoot them a text!* Exactly! Why can't you do the same thing with guests at your church? You can. You should. And . . . we'll help you. Phew, this gets me fired up.

Alright, but texting isn't the only way. We recommend emailing, writing a handwritten letter, making a personal phone call, even having a gift for them. Here, check out this simple follow-up calendar:

S	M	T	W	Th	F	Sa
S ✉	M f	T	W	Th	F ✉	Sa 💬
S ☕	M	T	W 📇📞	Th	F 💬	Sa
S	M	T	W	Th	F ⚡	Sa
S	M	T	W	Th	F ✉	Sa 💬
S	M	T	W	Th 💬	F	Sa ✉
S	M	T	W	Th	F ✉	Sa 💬

The envelopes represent emails, the Facebook and Messenger icons are for connecting with them on Facebook, the black text bubbles are text messages, the coffee cup is a "mug 'em" (Text In Church lingo for giving them a gift, like a coffee mug), the postcard is a handwritten note, and the phone represents a phone call!

We will walk through each of these, and I'll give you some examples. But first, I want you to notice how the consistency of the outreach changes. In those first two weeks, you are communicating a lot. This is intentional.

The very first time someone attends church, there's more than an 80 percent likelihood that they will never come back.

The odds are stacked against you. So, if you can love on them, send gentile reminders, encourage them over and over that week, why wouldn't you? It's not like you're hitting them up and asking them to give your church money. You are simply reaching out to say hello, to extend an invitation, or to just be thoughtful. But once they have started coming a couple of weeks in a row and maybe taken a next step by attending an activity, their "vulnerability," so to speak, of dropping off the map goes down significantly.

Once a guest has attended your church for three to four weeks, they hopefully know at least one or two familiar faces and have taken a next step, or at least have one on their calendar. At this point, they don't need to be nurtured like a first-time guest. And honestly, they don't want to be. Nobody wants to feel like the new kid, right? So the sooner we can connect with them in a meaningful way, the quicker trust is built and a new pattern is created where church becomes "normal." We aren't talking about integrating people into your church's body; we are simply talking about how to follow up with them in a way that is impactful and will keep them coming back.

Before we dive into the six critical and effective steps to following up with first-time guests, let's look at three huge mistakes churches are making.

THREE MISTAKES CHURCHES MAKE

Mistake #1: Inconsistency

Inconsistency is one of the biggest mistakes churches make when it comes to guest follow-up. When a first-time guest shows up at your church, more than likely, they have a lifestyle of not attending church. If invitations or introductions are sporadic and unintentional, they will get drowned out by the noise and busyness of life.

Consistent communication is key to showing a guest that they're important. We know you care, but they need to know you care. And the best way for them to know it is for you to show it.

Mistake #2: Irrelevancy

There are two ways your church can be irrelevant in terms of your guest follow-up:

1. Content
2. Method

Irrelevant content happens when the church sends the singles' ministry event invite to a married couple. Yikes. Don't do that. If you don't know anything about the person yet, the content you send them should be very basic: come to Sunday service, come to our lunch and learn, etc. As you get to know them better, you can send them more specific announcements that align with their interests and needs.

Irrelevant methods happen when a church ignores people's communication preferences or doesn't offer preferences at all. Don't just communicate exclusively via email because that's how you've always done it. Everyone prefers a different method of communication, so use a combination in order to stay relevant! For example, text, phone call, email, etc.

Mistake #3: Insufficiency

Being insufficient in guest follow-up communication is probably the most common mistake we see. And we totally get why. Sending multiple messages to several first-time guests week after week after week can become an overwhelming, time-intensive task.

However, insufficient follow-up is simply noise. People don't read quick email blasts. They feel automated and generic, and people see right through them.

Also, as we've talked about, many of your first-time guests have a lifelong pattern of not attending church. One or two follow-up messages isn't going to impact that.

When we talk about a sufficient follow-up process, we are talking about at least six weeks' worth of follow-up that includes text messages, emails, a phone call, a handwritten letter, and a gift. It's comprehensive and focuses on relationship-building. As you saw in the calendar, we aren't talking about sending texts and emails every day. Rather, we suggest reaching out with simple invites, check-ins, and reminders. Sufficient guest follow-up is investing in that person and doing your best to build a relationship with them.

Alright, now let's dive into the six follow-up strategies.

SIX FOLLOW-UP STRATEGIES

 *1. Text Them*

We all text. Even my grandmother texts. And 98 percent of those texts are read, so if you skip out on this one, you are missing one of the biggest opportunities to engage with first-time guests. The content doesn't need to be mind-blowing or formal. Just a simple "thanks for coming" or "I hope to see you tomorrow" goes a long way. We have actually found that the more casual and conversational these texts are, the better! People appreciate feeling like a pastor just had them on their mind and shot them a quick text. It creates a sense of intimacy similar to a friendship. Plus, the guest doesn't have to do the pursuing. They don't have to look up the service time again on the website. They have someone on the inside saying, "We want you here; we're thinking of you."

We recommend sending a text message on Friday or Saturday for the first three weeks after a guest visits your church for the first time. This will keep your church service front of mind as people make their weekend plans.

Here's a sample script:

"Hi James! We hope to see you tomorrow @ City Church! —Pastor Jason"

See? It's not really even a script; it's one line, and it's so simple.

2. Email Them

Send one or two emails each week for six weeks after a first-time guest visits your church. Thank them for taking time out of their weekend to worship with you. Give them links to your website, Facebook page, or podcast. And let them know you are just an email away if they have any questions!

As a quick aside, make sure you personalize this email. It's okay to start with a template, but add their name if you have it and use language that's informal and conversational. Connection is the goal, so your email should reflect that.

"Hey James! Pastor Jason here. I'm so glad you tried out City Church on Sunday. I hope to see you again next weekend and get the opportunity to meet you!

You can learn more about our church on our website _____, on Facebook @_____, and via our podcast _____!

Take care,
Jason Jones, City Church

Sending an email once a week for six consecutive weeks is not obnoxious. It shows that you want to stay connected and that you are intentional. These emails should invite first-time guests back to church or to an informal gathering where they can learn more about your church. Maybe it's a small group setting or a lunch and learn environment. If you don't have anything like this, I'd encourage you to consider starting it. Have a place where guests can get more familiar with your church outside of the corporate worship setting. And make sure you're emailing them to tell them how much you want them to be there!

3. Mug (or Gift) Them

Everyone likes free stuff! Even if "gifts" aren't your love language, you most likely wouldn't say no to a free gift from a church you were visiting for the first time. Right? Most people would appreciate this gesture of thoughtfulness. Whether it's a coffee mug, a Starbucks gift card, or even a pen, having something for first-time guests to take home with them reminds them they are valued and appreciated. And if you can afford to brand your gifts, this can make an even greater impact. Guests might just use your coffee cup during the week and think, *Maybe I'll give that church another visit.*

To be sure guests receive their free gift, you can either ask them to head to the Welcome Center or you can request their address on the physical or digital Connect Card and then drop the gift off at their house. Just don't be a lurker and invite yourself in! Again, the gift doesn't have to be expensive, and you only need to give one per guest/family. We aren't bribing them to come to church. It's just a simple token of appreciation, and you'll be surprised how far it goes.

4. Call Them

After a first-time guest has shown up at your church, someone from your church needs to give them a call within the first week. Thank them for coming to your church, let them know you are there if they have any questions, and ask them if there is anything that could have made their experience better. You don't need to stay on the phone for a long time unless they are continuing the conversation! Just be clear and quick. Take a look at this example:

Hello, _____! This is Tyler from City Church. I just wanted to call and thank you for joining us on Sunday! I wanted to see if you had any questions and invite you to join us again this coming weekend.

Now, I know what you are thinking: No one calls anymore. I can skip this one! And you're right. Most people, especially younger generations, don't like to talk on the phone anymore. But I urge you—don't skip this step! It is likely the person you're calling will not have your number and, therefore, won't answer the phone. If this happens, you can still leave them a super thoughtful voicemail.

5. Facebook Them

We all know that Facebook is popular. We also know at least a few people who overuse Facebook—come on, you know you just thought of a specific friend and chuckled a little bit. However, I don't think we understand just *how big* Facebook really is. So, allow me to blow your mind a bit.

- As of 2019, a quarter of the world's population is on Facebook. That's one quarter of the whole world! Almost two billion people!
- Americans spend fifty-eight minutes per day on Facebook.
- Facebook messenger has been ranked as the number one mobile app.
- Facebook is now ranked the third most visited website worldwide.
- Two hundred million people are members of meaningful Facebook groups.[2]

Clearly, Facebook is a place where people access information and connect with each other! Using Facebook thoughtfully as part of your first-time guest follow-up process shows that you are relevant and intentional.

Within the first week of someone visiting your church for the first time, someone on your church staff should send them a friend request on Facebook. Ideally, this would be from a staff member who they've met or who would be in contact with them regularly (i.e. the pastor of adult discipleship, the youth pastor, the children's pastor

---

[2] Paige Cooper, "41 Facebook Stats That Matter to Marketers in 2019," Hootsuite, November 13, 2018, https://blog.hootsuite.com/facebook-statistics/.

if they have kids, etc.). Then, after their second or third week, invite them to join one of your groups on Facebook or to like your page. This will keep your content in their feed and hopefully get them connecting with others from your church.

Facebook is an especially effective way to follow up if people haven't left any contact information. As long as you know their name, you can send them a message inviting them back the next Sunday!

Here's a sample Facebook connection script:

Happy Friday, James! I just wanted to send a quick invite to City Church this Sunday. We have many awesome ways to connect with others at City Church, and I'd love to help you get connected!

You can find our Facebook page here: [FB PAGE]

Hope to see you there!
—Pastor Jason

 **6. Mail Them**

You might be thinking, *Why's the texting guy telling me to write a note?* Well, because with the increasing amount of media that comes our way through technology, a handwritten note goes a long way. People appreciate the time and thought it takes to write a personalized letter, especially when they are on the receiving end. This shouldn't be a stamped, generic thank-you card but a *real* thank-you note written with pen on paper. In the first week or two of a guest visiting, send them a short, handwritten thank-you card in the mail.

Here's an idea of what it can say:

Hi, James! Thank you for joining us on Sunday at City Church! We would love to see you again this weekend. Please let us know if you have any questions or anything we can pray for you about. Feel free to call me at my number below at any time.

Have a great rest of your week!

Jason Jones, City Church
866-256-2480

See? Short and sweet, right? You have acknowledged that they visited your church, you have made yourself available for any questions or prayer requests they might have, and you've invited them back for another visit. That's all it takes!

These six strategies came out of my work with thousands of pastors across the country. If they can work for thousands of other churches, I'm confident they will work for your church too! Just use the calendar I shared earlier in this chapter and the six different ways of connecting, and you will have a successful follow-up plan that helps you connect with first-time guests and make them feel seen and wanted.

If you'd like to have free access to our Ultimate Guest Follow-Up Plan downloadable PDF from our Resource Library, just go to the Resources tab in the Boomerang Kit. Text kit to 816-482-3337 if you still need to get the link.

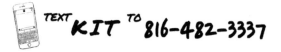

The one argument I hear against some of these strategies is that they're impersonal. People fear that technology (texting, emailing, social media) will replace face-to-face, human interaction. What I want church leaders to understand is that what we are doing with these tools is utilizing them to facilitate *more* human interaction.

Imagine this: You have a friend you haven't seen in years. You recently learned that this friend's mother has died. You want to connect with your friend, share your condolences, and just be supportive. So how do you go about doing that? You don't *hope* that you just run into them so you can do it in person. And I'm guessing you won't just show up at your friend's house. No. My guess is, you'd call, maybe send a text. You'd ask if you can get together, and you'd tell them they have been on your mind. They'd appreciate you reaching out so much that they'd agree to lunch the next week. Boom. You now have an opportunity to connect with this friend face to face, but it might never have happened without the use of technology.

Holly Howard, a church leader we work with, said this about these strategies:

> *We find out personal things from our new guests that help us connect to them and know how to effectively minister to them. They are surprisingly "open" with personal issues when we text or make a phone call. They have "spilled their guts"—and we love that! Instead of taking months or years to open up, they feel very free to share what they are going through after just a couple of texts or emails! Text In Church is speeding up the process of "real" ministry in their lives.*
>
> *When we first started using Text In Church, we added 19 people through a Connect Card. Of that 19 people, 11 returned. That's a 58% return rate! That's up from around 5% for us! And of those 11, they all took significant next steps, like completing our Connection Point class, getting involved in a Dream Team (serving), giving, and dedicating their lives to Jesus!*

Technology can be a powerful ministry tool! Our passion at Text In Church is to help church leaders use it to its maximum potential.

A FOLLOW-UP PLAN THAT WORKS SUMMARY

There are three mistakes you must avoid when it comes to follow-up:

1. Inconsistency: communication that is random and unintentional
2. Irrelevancy: method and/or content that isn't relevant to the person
3. Insufficiency: a lack of communication

There are six proven follow-up strategies you can utilize over at least six weeks to make sure that your first-time guests feel known, noticed, and loved.

💬 Text Them

✉ Email Them

☕ Mug (or Gift) Them

📞 Call Them

f 💬 Facebook Them

📇 Mail Them

When a first-time guest shows up at your church and leaves their contact information, it is our great privilege and responsibility to nurture that relationship. While it does need to be intentional and thoughtful, it doesn't need to be complicated. First-time guests are people seeking community, and no one creates that better than the church!

DISCUSSION QUESTIONS

What is your church's follow-up strategy?

Are you making any of the three mistakes when it comes to your follow-up strategy?

Which of these six follow-up steps are you currently doing?

Which ones can you implement ASAP? Which ones would you like to implement over the next six months? Who has the capacity to head this up at your church?

Is your follow-up process fostering relationship-building? Are you hearing back from first-time guests?

BUILD

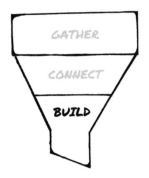

I am in charge of guests' experiences at my church—from Plan Your Visit/First Contact through the six-week guest follow-up process. We have been doing Plan Your Visit for nine months and a six-week follow-up for two months. We were using our own systems and methods like Mail Chimp, CCB Mail Merge, and regular texts. I had been old-school texting every guest for six weeks, and we have ten to twenty guests every week. So, I was sending sixty to one hundred and twenty individual, personal texts plus replies every single week! The time! Let's just say my world was spinning out of control.

Then, my lead pastor saw a Text In Church email and told me to check out a webinar. Text In Church affirmed what I'd been telling my church: texts get opened more often. I tried out Text In Church by putting my regular process into an automated flow and sending all six weeks of info to my pastors super quick (in one day). They approved and—hallelujah!—my job got a whole lot easier. All I have to do is put my guests into my system and then watch them respond to me throughout the week. The guests who don't want to talk to us can still be in the system via the digital Connect Card if they choose. We also traded out our previous Plan Your Visit form for a digital card. Again—amazing!

My Plan Your Visit guests automatically get a text from me, asking which service they would like to plan to visit (which opens our dialogue for pre-registering kids, etc.), followed by a detailed email from me, and finally a text telling them I can't wait to see them.

—Faith Brown, Crosspoint Fellowship Church

As I mentioned in the Introduction, this idea of churches using text messages was inspired by a pretty horrible night. I won't rehash the whole thing for you, but the highlights for those of you non-intro-readers are:

- The church plant I was a part of came up with a stellar follow-up plan.
- Nobody on the leadership team wanted or had the time to execute it.
- We all went home defeated and frustrated.
- I got a text from Southwest Airlines telling me my flight the next day was already delayed.
- I got more frustrated.
- I had a light-bulb moment!

After getting the text from Southwest, I thought to myself, *If an airline can get in touch with complete strangers using text messaging, why can't the church use it to communicate with guests?*

There were lots of kinks to work out. My brother, who was planting a church in California at the time, had some hard pushback when I ran the idea by him. He said Southwest was using text messaging as a megaphone. Their intention wasn't to get responses back. They simply wanted to spread their message. Churches, on the other hand, are all about relationship-building. There's no way they can send these announcement messages from a five-digit number that feels like a marketing message from Pizza Hut and make any sort of impact.

He was right.

So I went back to the drawing board.

Texting for churches didn't have to look exactly like it did for Southwest. However, Southwest did have something to their strategy that I couldn't let go of. When my flight got delayed, there was no way a real person went through each passenger on that flight and texted them individually. No, a system did that. It was automated, and no one at Southwest had to think about it.

So, through a lot of trial and error, we merged the familiarity and personalization of a text message with the efficiency of automation and built something really special.

That is what I encourage you to do with all of these strategies. Part of why I included summaries and discussion questions in this book is so that you and your team can really dissect this content and figure out what your church needs to start doing now, what you can build and implement over the long haul, and how you can use systems to do a lot of the work for you.

CHAPTER 10

BUILDING A TEXTING STRATEGY

Texting has always seemed like a natural communication channel for me. On the other hand, I weed through my email more harshly than probably anyone else. My social media is always full of notifications I don't check. I never answer a phone call from a number I don't know.

But a text?

That gets my attention.

And I'm not alone. Most people prefer texting. When I began researching this years ago, the numbers were staggering. But today they are even bigger:

- Twenty-three billion texts are sent every day. That's *billion*![3]

---

[3] Teodora Dobrilova, "35 Must-Know SMS Marketing Statistics in 2019," TechJury, May 13, 2019, https://techjury.net/stats-about/sms-marketing-statistics/#gref.

- Ninety percent of texts are read within three minutes. Email open rates are at around 20 percent, and it can take hours to days before they're opened.[4]
- Forty-five percent of texts are responded to. There is absolutely no other communication method that even comes close to this response rate.[5]
- And, finally, Ninety One percent of phone users keep their phone within three feet of them twenty-four hours per day.[6]

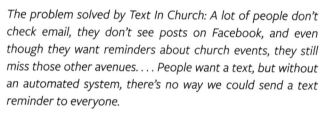

The problem solved by Text In Church: A lot of people don't check email, they don't see posts on Facebook, and even though they want reminders about church events, they still miss those other avenues. . . . People want a text, but without an automated system, there's no way we could send a text reminder to everyone.

—Teddi Deppner, Lincoln Christian Life Center

Now, stop reading. Where is your phone? Is it within arm's reach? My guess is, it is—just like mine. The proof

---

[4] "Conversational Advertising," Mobile Squared, https://mobilesquared.co.uk/wp-content/uploads/2017/12/Conversational-Advertising.pdf.

[5] Velocify, https://www.velocify.com/blog/text-me-maybe-how-to-use-automated-text-messaging-to-win-love-leads-revenue.

[6] Cheryl Conner, "Fifty Essential Mobile Marketing Facts," Forbes, November 12, 2013, https://www.forbes.com/sites/cherylsnappconner/2013/11/12/fifty-essential-mobile-marketing-facts/#7fedd9087475.

is in the pudding. Texting is the number one way people communicate, yet less than a decade ago, churches were not seen using it!

Now, I've never believed that texting should be the only communication tool you use. Obviously, you read in the Follow-Up Plan chapter that texting is just one of six critical steps to following up with first-time guests. But I can't stress enough its importance. To try to communicate with your first-time guests without the use of text messaging is like trying to corral a packed house at Kauffman Stadium (home of my team, the Kansas City Royals. Go Royals!) without a microphone. You might get one or two people's attention, but no one else will even notice you're talking to them.

> TO TRY AND COMMUNICATE WITH YOUR FIRST-TIME GUESTS WITHOUT THE USE OF TEXT MESSAGING IS LIKE TRYING TO CORRAL A PACKED HOUSE AT KAUFFMAN STADIUM WITHOUT A MICROPHONE. YOU MIGHT GET ONE OR TWO PEOPLE'S ATTENTION, BUT NO ONE ELSE WILL EVEN NOTICE YOU'RE TALKING TO THEM.

Also, you need to understand that a church that texts and a church that has a texting strategy are two very different things. I am not condoning sending a text out every morning with your favorite Bible verse and a list of activities happening at the church.

You do you, but I don't recommend it.

I am talking about using texting as part of a very well thought out, personalized strategy to engage your first-time guests. We've already gone over the process in the Follow-Up Plan chapter, and we will walk through the "how" in the next two chapters, but here are a few tips specific to texting.

1. You'll need a local phone number.

Use a ten-digit, local phone number from which to text your first-time guests. This humanizes the text, rather than sending texts from a short code (a five-digit number) that people immediately write off as spam. It also allows for two-way messaging. This is how my brother and I got away from the Southwest megaphone dilemma. Churches don't want to use texting to shout information at people; they want to build relationships and foster connection. Two-way messaging allows you to send texts back and forth with another individual.

2. A short code is a must-have.

Again, short codes are five-digit numbers, exactly like the number that sent me my notification from Southwest. Have you ever received an appointment reminder from a doctor's office? Most likely it, too, came from a short code. Short codes get a lot of grief. However, they can ensure 100 percent deliverability. The best-case scenario is that your church can communicate using a ten-digit phone number for all of its correspondence. But, unfortunately, with the way carrier filtering has continued to change and impact the churches we work with, we have found supplemental short code capability extremely important. The long and short of it is, a cell phone carrier gets to filter messages however it deems appropriate. So, if you try to send a link from a ten-digit phone number, for example, the carrier might flag the message as spam, and it wouldn't get sent. In this case, you can simply send any message with a link through your short code and then use your ten-digit number for correspondence.

If you are starting to feel a little overwhelmed by all this technology talk, don't worry. I know a fantastic system that will do all this for you.

3. Finally, text like you text a friend.

When you're using texting as a part of your follow-up strategy, remember that you're texting another human being. This is your only opportunity for personal conversation until you can get face to face with guests, so be friendly and have some personality! Don't sound robotic and formal. People don't like texts that feel automated or computer generated. So just pretend you're texting a friend! Hopefully, it won't be too long before that's the case anyway.

BUILDING A TEXTING STRATEGY SUMMARY

Texting is a critical tool your church needs to be using. The proof is in the numbers:

- Twenty-three billion text messages are sent each day.[7]
- Ninety percent of texts are read within three minutes.[8]
- Forty-five percent of texts are responded to.[9]
- Ninety-one percent of phone users keep their phone within three feet of them twenty-four hours a day.[10]

Whether you use a system like Text In Church or personal cell phones, make sure you follow these three texting rules:

1. Use a local, ten-digit phone number; this feels personal to people and allows for two-way messaging.
2. Use a short code; this ensures 100 percent deliverability.
3. Text with personality—like you are talking to a friend!

---

[7] Teodora Dobrilova, "35 Must-Know SMS Marketing Statistics in 2019," TechJury, May 13, 2019, https://techjury.net/stats-about/sms-marketing-statistics/#gref.

[8] "Conversational Advertising," Mobile Squared, https://mobilesquared.co.uk/wp-content/uploads/2017/12/Conversational-Advertising.pdf.

[9] Velocify, https://www.velocify.com/blog/text-me-maybe-how-to-use-automated-text-messaging-to-win-love-leads-revenue.

[10] Cheryl Conner, "Fifty Essential Mobile Marketing Facts," Forbes, November 12, 2013, https://www.forbes.com/sites/cherylsnappconner/2013/11/12/fifty-essential-mobile-marketing-facts/#7fedd9087475.

DISCUSSION QUESTIONS

Is your church currently using text messaging to follow up with first-time guests?

If not, why?

If so, how are you using texting? How often do you text?

Do you have the three texting rules covered?

Who has time to manage this part of the follow-up process? What phone number will they use?

CHAPTER 11

AUTOMATION

All of the strategies and technology tools we've talked about so far are great and exciting, but you might be thinking, *How am I supposed to pull this off?* People working at a church aren't exactly known for having light, open schedules. Their plates are full and the processes they have in place—or don't have in place—are there for a reason. Am I right?

Look, that's the exact same brick wall I hit years ago when I was trying to grow a church plant. Time and money are in short supply. I didn't have time to connect with every single guest on a Sunday morning, much less reach out to them with various communication methods over the span of six weeks.

However, something was happening at the same time. The concept of digital marketing was just starting to take off. Companies had figured out how to schedule emails to go out on a certain day at a specific time. Businesses realized that more people were shopping online and reading their emails than shopping at malls and reading store mailers. There was a shift happening, and as different as it was, it was exciting!

I started to introduce this idea of automation to the pastors I was working with back then. However, I knew we needed a system that was more comprehensive than simply delivering an email every Thursday. We actually looked at what successful businesses were doing and decided to see if any of it could translate over to help the church. We ended up modeling a lot of what we do at Text In Church after Infusionsoft, the premier marketing automation platform for small- and medium-sized businesses.

> Our church implemented the guest follow-up automated workflows, and someone who filled out the digital Connect Card started receiving these messages. She told our church this was the first time she ever felt wanted and noticed from a church. It was a super simple gesture, but to her after not getting any communication from the last church they visited, even after they had been gone for a couple of months, it felt like a huge touchpoint just getting a simple text. Now they are regular attendees and are getting involved in helping plant another church. Talk about a cool story!
>
> **—Trinity Reformed Church**

What this led to was the introduction of automated workflows into the church space. Automated workflows are a series of messages created to be sent to a designated group of people triggered by their joining a group.

For example, let's say Anna visited your church for the first time on Sunday. She texted in your keyword "Welcome," signaling that she's new, and she filled out the Smart Connect Card on her phone. Within an automated system like Text In Church, her contact information is then directly added to your account and triggers the automated workflow setup for first-time guests. This includes six weeks' worth of emails, texts, and reminder messages to staff to make a phone call, write a note, or send a friend request on Facebook.

Automation can and should lead to personal interactions, and that's exactly what we've seen.

If you use multiple types of Connect Cards, you don't know when someone will be added to your system. Automation makes sure they get the appropriate message at the exact right time, regardless. So, they won't get invited to an event that happened two days ago!

You have designated work hours (I hope) and shouldn't have to be available for follow-up 24/7. Automation takes care of it, making sure a text goes out Saturday morning without you having to think about it. If technology can do the work for you, why wouldn't you let it? Save yourself some time and sanity.

Some of the best times to reach out to people are not ideal times for you to be sitting at your desk cranking out texts and emails. Automation allows you to set messages up to send at those ideal times, while you get to focus on other areas of your ministry.

One of our members described this so beautifully. He said:

> *Text In Church enables me to create the relationship and ministry footprint that both guests and volunteers need. I couldn't make this kind of contact without you guys. My volunteer show-up rate has drastically improved. I use Planning Center for scheduling. This is relational! Guests are coming back! I have led two guests to start relationships with Jesus this week, and they are joining small groups and serving. I know about the need to connect, so this guest piece was happening before Text In Church. The problem was, I couldn't do anything else. Now I'm able to do more ministry and help in other ways also! It's just amazing!*

Automation is really the key to what we do. Automation is what allows the reach of one church to far exceed what a person can do. Automation is what helps church leaders avoid burnout, because tasks that have taken them hours and are a high priority can now be done in a matter of minutes. Automation is what allows someone outside of the church to feel virtually embraced and more comfortable walking through the doors again because they know someone has thought of them. Automation is one of technology's gifts for connection, and, when used correctly, it will make a huge impact on your community.

Text In Church has allowed us to reach out to our first-time guests within a few hours of them coming and ask how we can connect them. Our initial text includes a welcome and asks if they have any questions. The responses have been amazing. About 50 percent of the people we send a text to after their first visit respond back to us. The responses range from "no questions at this time, but thanks for reaching out" to "I am looking for more information on a small group/women's ministry/children's programs/baptism/etc." Then, there are some really poignant ones like, "my wife just passed away. Do you offer grief groups?" or one from a mom who has terminal cancer and wants to help her family get connected at a church before she passes. People also share their experiences of our church. Some say, "It was a beautiful service," and we often here things like, "I had tears streaming down my face." These are contacts that normally would take us much longer to make if not for this invaluable tool.

—Melanie Hill, Church of the Resurrection

Being a pastor or church leader is definitely a relational calling. And even hearing the word automation can make you cringe because it usually feels robotic, impersonal, spammy even. But you only feel that way because the content of the message is all wrong.

What if automation could save you time and ensure

your personalized follow-up messages were sent whether you were at work, at home sick, or on vacation? What if automation could allow your church to far exceed the work of any one person and reach more people? Obviously, there are some negative things about automation. I'm sure we've all been on the receiving end of spam. But we avoid this by personalizing all of the messages. Texts and emails can include the recipient's name, the sender's name, your church's logo, and normal, everyday language.

Additionally, automated messages can still use two-way messaging. You don't have to worry about an automated response being sent when someone responds with something like, "thanks, but we will be out of town for a funeral," and you now have some really important information this person shared with you. You know how to be praying for them, you can schedule a meal to be delivered to their family, or you can simply respond telling them you're so sorry for their loss and that you'll be thinking of them. This is how relationships are built in the twenty-first century. And I can promise you, this makes a huge impact on people!

AUTOMATION SUMMARY

Automation leads to personalization and creates huge opportunities to connect with more people.

Automated workflows are a series of messages created

and scheduled to be sent based on when someone joins a group. This can happen automatically from a Smart Connect Card or once someone is entered into the database by an admin.

Big NOs for automation are generic and megaphone-style messages. Personalize your messages so that people don't feel like they're hearing from a robot or that it could have been sent to anyone. Make sure they know this text is meant for *them*, that you thought about them specifically. Additionally, make sure the goal of your messages is connection; ask a question or send some sort of invitation. Don't just "yell" information at them.

DISCUSSION QUESTIONS

Does your church have any automated processes?

How has automation benefitted specific areas of your church?

How can you implement automation into your guest follow-up system?

CHAPTER 12

A SUSTAINABLE SYSTEM

Well, there you have it. Every "secret" I've learned from working with church leaders is laid out for you. I want church leaders and communicators everywhere to be equipped with as many tools in their tool belt as possible. So, we are literally an open book at Text In Church. I don't think having a system like the one our team has built is necessary to be intentional and connected. However, I do think it's necessary in order to be sustainable. Even if you're not a systems geek like I am, you can't deny the impact they have. Let me give you some context for why I believe that.

In 2019 alone, Text In Church members sent over 3,200,000 emails and over 25,300,000 text messages.

Go back and read those numbers again.

That is over *twenty-eight* million emails and texts! And in case you can get a little data-driven and forget what an email and text really are, those are *people*! Over twenty-eight million times churches reached out to *people* in their communities, inviting them back, encouraging them, praying for them, thanking them for serving.

> *We have been successfully utilizing Text In Church for an eight-week automated workflow for guest follow-up, sending about one text or email per week. It has been very well-received and has proved to greatly help retention! It makes it personal for each guest and allows them to be reminded of service times, ask a question, or hear about an event. Texts are always read with a high engagement rate when I ask a question. We would never be able to do that without this service!*
>
> **—Danielle Bettmann, Brookside Church**

If you've struggled with the use of technology in the church, these numbers alone should help to alleviate some of that fear. That is a reach far outside of our capability as individuals.

I was at a conference recently where a presenter was talking about church systems and about their sustainability. She posed the question: "If your church doubled in size at midnight tonight, would the systems you have in place today still work tomorrow?"

How does that question sit with you? I know it was eye-opening for me to think about.

I intentionally left you discussion questions at the end of each chapter so you and your team can work through this content and figure out a way to implement what is most important for your church. If you choose to do this without any automation or a system, I have a few tips:

Create Reminders: It's impossible for one person to remember all of this for each and every guest. So, make lists, put reminders on your calendar, set alarms on your phone. Do whatever you need to do so that people don't fall through the cracks.

Ask for Help: Church leaders wear a lot of hats. Too many, if you ask me. So, delegate some of this to other staff or a volunteer. Maybe there is someone in your congregation who loves data entry and can enter all of the information from your paper Connect Cards into your database every week. Empower that person to use this passion of theirs to serve the church. Also, buy them coffee and compliment them all the time because these people are a rare breed and you don't want to lose them. Perhaps some of these tasks can be divided among the different ministry staff. For example, the children's director can handle all the new families with kids, the youth pastor can handle families with middle and high school kids, the music minister can handle the sixty-five-plus age group if you don't have a pastor dedicated to those folks, and so on. If you're a staff of one or two, lean into your volunteers and leadership. It's amazing what people can do when they're empowered. Good leadership seeks to multiply, so equip someone to take ownership over a piece of this. For example, who attends your church and is passionate about first-time guests? Maybe they wouldn't use those words to describe themselves, but they're always out in the foyer before and after service talking with people. They invite friends and family to church all the time. They are just energized by the interpersonal interactions. This is your hospitality person! Put them in charge of Sunday morning in terms of greeters or hosting the people who planned a visit from your website. This person can then create their own team of volunteers to help execute this well, and you can focus on preaching or kids' ministry or whatever it is that you feel pulled away from.

Delegation is a powerful tool for effective leaders, not a sign of weakness. Don't let your pride get in the way!

PICK AND CHOOSE

Pick and Choose: Part of the reason I have you work through each of these strategies separately is so that you don't try to take an "all-in" manual approach. Start with the most important strategy or two and work to perfect a system for that/those. Once you feel confident with that strategy and you know you can sustain it, work to add in another strategy. If you try to do it all at once, the likelihood that it won't be done thoroughly—or worse, that people fall through the cracks—is much greater.

However, if you are considering a system like Text In Church, here's what I want you to know:

You see, Text In Church isn't just *any* church texting service. We didn't try to find a hole in a market so that we could fill a need and make a profit.

Far from it.

We were started when I was running my own photography business and had a brand-new family— too busy to head up the guest follow-up initiative at my own church.

We were started when church was losing its place in a family's weekend plans, when more people were at soccer games on a Sunday morning than at church.

We were started when text messages were becoming the best way for businesses to connect in a quick and automated way with their people, and we thought, *Man, what if we could utilize this technology to connect with people at church?*

We were started at the peak of the mobile movement, when more people were accessing the internet, and each other, through their phones than ever before.

We were started at a time when individuals raised in church were raising their families without church.

We were started when church leaders were struggling to find time to effectively make connections with the guests who walked through their doors.

We were started because we deeply believe the world needs the Church. And Text In Church gave all of us techy folks a place to use our skills and experience to do ministry, helping local churches leverage technology to reach more people. It really is that simple for us. And sticking to this mission as a company keeps us connected to a much greater mission that we are thankful every day to be a part of—growing the Kingdom of God.

Now, we are also really legit! Everything we've discussed in this book, Text In Church can do for you. We can schedule, automate, and customize just about anything you can dream up when it comes to marrying technology with guest follow-up. But don't take our word for it. It might be more powerful if you hear it from other church leaders:

Text In Church is very user-friendly and has made communicating with our teams and congregation so much more efficient. I highly recommend them! I love being able to send mass texts to people, and when they respond, it just comes back to me and not a huge thread of people. That's huge! I love being able to put together several emails and texts in one sitting and schedule to send them out later as needed. The fee is very reasonable—it's like having an assistant for $368 for the whole year! The staff is incredibly helpful, and I feel like they are committed to making us successful! Using Text In Church has freed me from spending so much time on my computer and has allowed me to spend more time with people! I highly recommend them!

—Cathy Lynn Morgan, Bethany Bible Church

I cannot endorse Text In Church and the services they provide enough. It has really revolutionized the way I've been able to contact and stay in contact with my people. Here are a few things I want to share: #1: You gotta text. That's the way people communicate today, and Text In Church allows you to do that. #2: Not only is it texting, but it's personal. It gives you a personal, local number so people don't feel like they're getting a group text, and you can actually personalize the text and put their name in it, so that's money right there. #3: It's just easy! Do yourself a favor and sign up for this service right now; you will not regret it.

—Jeff Johnson, Florence Baptist Temple

We use Text In Church, and it has been amazing! We run 1,100-ish on Sundays, so I use it for a six-week guest follow-up plan. I send between sixty to 120 texts every week to guests, plus an equal number of emails. I also use it to make contact with about 150 First Impressions volunteers every week. I just started using it this week to make contact with about fifty Journey Team leaders also. The response has been phenomenal. We started a six-week follow-up this February and have had over 50 percent of first-time guests take next steps (become regulars, commit to membership, receive salvation, join small groups, participate in baptism, etc.). Text In Church makes this manageable for me. Sending all the individual messages was killing my time, and I wasn't about to go all impersonal with other automated systems. I send messages to people, hoping to create dialogue. The first week I did this, I got lots of prayer requests. Then last week I asked people why they love to serve. My intent was to develop relationships since I was relatively new to the staff. What happened was, I had the best volunteer turnout ever! Text In Church is a big YES from me!

—Faith Brown, Crosspoint Fellowship Church

The reason Text In Church is making such an impact is because the platform was literally inspired by and built for churches. And it still exists for that purpose! So, when a church leader comes to us and says their needs are shifting, can we make changes to the platform, we look into it. We are always adapting and improving so that we can provide the very best tool to churches.

CONCLUSION

Well, there you have it. A complete culmination of what my team and thousands of church leaders have collaborated on. Technology is a funny thing, isn't it? It can create more unity and division than anything else I've witnessed in my lifetime. I'm on a mission to make sure that it does a much better job of creating unity. I want churches equipped with every technology tool possible. No longer can society call us irrelevant, ancient, and out of touch. Churches have the greatest news on the face of the earth, and I love being a catalyst for getting that news to travel as far and as often as churches can with the tools we have in front of us.

I've always been an entrepreneur. I started my first business when I was a teenager. Working and innovating are in my blood. But I've never had a passion for my work like I do now. And I believe what continues to fuel that is working alongside church leaders like you, who are showing up, day after day, fueled by this passion for the Kingdom of God. I pray this book is helpful, that these strategies are impactful, and that, as a result, your church extends its reach to those in the community who have felt unseen, intimidated, or even uninterested.

At Text In Church, we meet as a staff every Monday. I begin each meeting with our purpose: We exist to help church leaders leverage technology to build relationships and connect with their communities. That's our heart behind this book. Use it. Sit with your team and talk through these strategies. Grab a whiteboard and build out some of these concepts. Ask the people in your church to weigh in when you feel like you're so in the trenches you can't see it anymore. But, most of all, be encouraged. We are with you, we are champions for the local church, and we will continue to build the best system we can to equip you!

ABOUT THE AUTHORS

Tyler Smith, a married father of two from Kansas City, began his career as an entrepreneur at a very young age. As a 17-year-old, he went into business with his dad doing photography for boys' baseball tournaments. However, as he got older and started his own family, his passion was pulling him more and more into the local church.

Working alongside his brother (a United Methodist Pastor), as well as serving on the Leadership team for his home church, Tyler was bombarded with the time and noise trap so many church leaders find themselves in.

Taking what he'd learned from his first business, Tyler began applying some of these business principles to his and his brother's churches. Soon, word began to spread and Text In Church was born.

Ali Hofmeyer is the Content Manager for Text In Church. Her role is the perfect blend of her passions: writing and the local church. Ali and her family are very involved at their home church and have seen how the strategies and systems taught in this book can have a real, tangible impact on the lives of families and individuals. Co-writing Boomerang has been a highlight of Ali's career, as she's been able to pour the heart of her company on to the pages of this book. She hopes this will create awareness and empowerment for churches everywhere to tap into the technology at their fingertips to grow their reach!